Advanced Textbooks in Control and Signal Processing

Springer
London
Berlin
Heidelberg
New York
Barcelona
Hong Kong
Milan
Paris
Singapore
Tokyo

Series Editors

Professor Michael J. Grimble, Professor of Industrial Systems and Director
Professor Michael A. Johnson, Professor of Control Systems and Deputy Director
Industrial Control Centre, Department of Electronic and Electrical Engineering,
University of Strathclyde, Graham Hills Building, 50 George Street, Glasgow G1 1QE, U.K.

Other titles published in this series:

Genetic Algorithms
K.F. Man, K.S. Tang and S. Kwong

Model Predictive Control
E.F. Camacho and C. Bordons

Introduction to Optimal Estimation
E.W. Kamen and J. Su

Discrete-time Signal Processing
D. Williamson

Neural Networks for Modelling and Control of Dynamic Systems
M. Nørgaard, O. Ravn, L.K. Hansen and N.K. Poulsen

Modelling and Control of Robot Manipulators (2nd Edition)
L. Sciavicco and B. Siciliano

Fault Detection and Diagnosis in Industrial Systems
L.H. Chiang, E.L. Russell and R.D. Braatz

Statistical Signal Processing
T. Chonavel and S. Vaton
Publication Due April 2001

L. Fortuna, G. Rizzotto, M. Lavorgna,
G. Nunnari, M.G. Xibilia and R. Caponetto

Soft Computing

New Trends and Applications

With 193 Figures

Springer

Professor Luigi Fortuna
Dipartimento Elettrico Elettronico e Sistemistico, Università di Catania, Viale A. Doria 6, 95125 Catania, Italy

Doctor Gianguido Rizzotto
Doctor Mario Lavorgna
STMicroelectronics, Stradale Primosole 50, 95121 Catania, Italy

Professor Giuseppe Nunnari
Professor M. Gabriella Xibilia
Doctor Riccardo Caponetto
Dipartimento Elettrico Elettronico e Sistemistico, Università di Catania, Viale A. Doria 6, 95125 Catania, Italy

ISSN 1439-2232
ISBN 1-85233-308-1 Springer-Verlag London Berlin Heidelberg

British Library Cataloguing in Publication Data
Soft Computing : new trends and applications. - (Advanced Textbooks in control and signal processing)
1.Soft computing
I.Fortuna, L.
006.3
ISBN 1852333081

Library of Congress Cataloging-in-Publication Data
Soft computing : new trends and applications / L. Fortuna ... [et al.].
p. cm. — (Advanced textbooks in control and signal processing)
Includes bibliographical references.
ISBN 1-85233-308-1 (alk. paper)
1. Soft computing. I. Fortuna, L. (Luigi), 1953- II. Series.
QA76.9.S63 S6334 2000
006.3—dc21 00-046342

Typesetting: Electronic text files prepared by authors
Printed and bound by Athenæum Press Ltd., Gateshead, Tyne & Wear
69/3830-543210 Printed on acid-free paper SPIN 10764606

Series Editors' Foreword

The topics of control engineering and signal processing continue to flourish and develop. In common with general scientific investigation, new ideas, concepts and interpretations emerge quite spontaneously and these are then discussed, used, discarded or subsumed into the prevailing subject paradigm. Sometimes these innovative concepts coalesce into a new sub-discipline within the broad subject tapestry of control and signal processing. This preliminary battle between old and new usually takes place at conferences, through the Internet and in the journals of the discipline. After a little more maturity has been acquired has been acquired by the new concepts then archival publication as a scientific or engineering monograph may occur.

A new concept in control and signal processing is known to have arrived when sufficient material has developed for the topic to be taught as a specialised tutorial workshop or as a course to undergraduates, graduates or industrial engineers. The *Advanced Textbooks in Control and Signal Processing* series is designed as a vehicle for the systematic presentation of course material for both popular and innovative topics in the discipline. It is hoped that prospective authors will welcome the opportunity to publish a structured presentation of either existing subject areas or some of the newer emerging control and signal processing technologies.

Fuzzy logic, neural networks and genetic algorithms have excited much interest among control engineering practitioners. This interest has arisen because these methods are able to solve a new set of control engineering and signal processing problems, which are more loosely defined and sometimes involve a degree of operational uncertainty. Fortuna *et al.* tackle the problems of image processing, greenhouse temperature control, urban traffic noise monitoring, a robot for picking citrus fruit, a rapid thermal processing oven and air pollution control in this new course textbook for the *Advanced Textbooks in Control and Signal Processing* series. This is a very interesting set of real applications problems and for the first time in the textbook series the emerging technologies have been treated as a solution paradigm for these problems. Thus the book first creates the pedagogical basis for the subject in early chapters on fuzzy logic, fuzzy control, artificial neural networks and so on, before progressing to the descriptions and solutions of the applications problems.

This textbook presentation of the emerging technologies should be invaluable for new courses in this area, since all the approaches are presented together. The textbook should also be useful for advanced undergraduate projects, and home-study by practising engineers.

M.J. Grimble and M.A. Johnson
Industrial Control Centre
Glasgow, Scotland, U.K.
October, 2000

Preface

Since the invention of the transistor in the 1947, the semiconductor industry has passed from manufacture of a single component to that of a device, containing more than a billion transistors, integrated in a few square centimeters of silicon.

In the last 25 years, there has been an annual reduction of 26 per cent in the price of a bit; nowadays, memory of more than a billion bits and teraflops (10^{12} floating point operations per second) processors are produced. This technological development cannot be compared with those in other industrial activities.

If the automotive industry had proceeded with the same speed, a car could cost about ten dollars, its top speed would be very close to light speed and with just a little bit more than a liter of fuel, it could be possible to cover ten times distance from the Earth to the Moon; so it is easy to guess how microelectronics has had such an important role in the process of welfare and the improvement of the quality of life and it is easy to believe that this will be the tendency in years to come. After this preliminary statement, soft computing's computational theory seems to be able to accept the challenge.

Soft computing's purpose is to keep up the techno-scientific project, started in the eighties with fuzzy logic, first in Japan and then in the West, where the complexity, the non-linearity, the uncertainty, intrinsic properties of real processes and phenomena, are no longer considered as a limit for the planning of the systems but on the contrary, as the starting-point for the study of new inferential mechanisms and non-conventional algorithms for the creation of more tolerant, robust and cheaper products.

In other words, soft computing is an antithesis of hard computing: it provides algorithms that are able to value, to reason and to discriminate, rather than just to 'calculate'; these new structures of calculation are based on logic with more values (not just true or false, but different degrees of certainty) and inspired by natural processes like selection, aggregation and co-operation.

At full speed, soft computing, together with the technology of integration, makes it possible, from today, to plan more intelligent and easier-to-use machines, increasing the quality of our daily life, in both social and work environments.

Thanks to scientific collaboration with a lot of universities, STMicroelectronics can be proud of a center of excellence for planning intelligent products, based on soft computing methodologies, that are involved in automotive and electronic systems, servo-systems for industrial applications, consumer products, and so on.

This book arises from the co-operation between the academic world, and in particular between the University of Catania and STMicroelectronics.

The authors synthesize the most important results of this decennial collaboration in a clear and complete way.

The bilateral industry-university results, are, once again, not just successful, but also, the only way to transfer innovative ideas from the research world to the industrial one, in a short time, making, the gift of soft computing methodologies at the disposal of society.

Salvatore Castorina
STMicroelectronics Corporate Vice-President

Table of Contents

1. Introduction

During the last decade, we have witnessed the evolution of some information processing techniques, not altogether of analytical type. In particular, there have been those of *fuzzy logic*, *artificial neural networks*, and, as well, the global optimization ones, such as *evolutionary algorithms* and *simulated annealing*.

The main ideas of these methodologies have their roots some way back in the past: the beginning of the 1960s with regard to artificial neural networks and evolutionist-type optimization algorithms, and the start of the 70s for fuzzy logic.

Recently, thanks also to low computation costs, we have seen the massive development of these strategies. Before the 90s, it was basically a parallel development, in the sense that it was hard for groups of researchers operating in any one of those sectors to be involved in the techniques being developed in some other.

At the beginning of the 90s, the need was first felt of integrating one methodology with another. In fact, it was understood that a *synergetic use* of the methodologies mentioned would lead to tools that were certainly more powerful than if the techniques were employed individually.

Lotfi Zadeh was the first to raise systematically the problem of integrating techniques peculiar to fuzzy logic with neural and evolutionist ones, and in 1992 he named that integration *soft computing*.

Soft computing is thus a methodology tending to fuse synergically the different aspects of fuzzy logic, neural networks, evolutionary algorithms, and non-linear distributed systems in such way as to define and implement hybrid systems (*neuro-fuzzy*, *fuzzy-genetic*, *fuzzy cellular neural networks*, *etc.*) which from one time to another manage to come up with innovative solutions in the various sectors of intelligent control, classification, and modeling and simulating complex non-linear dynamic systems.

The basic principle of soft computing is its combined use of these new computation techniques that allow it to achieve a higher tolerance level towards imprecision and approximation, and thereby new software/hardware products can be had at lower cost, which are robust and better integrated in the real world. The hybrid systems deriving from this combination of soft computing techniques are considered to be the new frontier of *artificial intelligence*.

In 1994, the classical IEEE meetings on fuzzy logic, neural networks and genetic algorithms that had been strategically held in different venues were for the first time brought together in one common congress in Orlando (USA) where it was officially shown that only an interaction of several different techniques could lead to problem solving when we are faced with uncertainties and partially structured models.

In recent years, it is not just the academic world, but above all industry, that has been focusing attention on this type of approach and has evidenced how soft computing techniques are advantageous and can be rapidly implemented in numerous sectors, such as pattern recognition, complex process control, and signal processing.

The constant attention paid to these techniques by the productive reality has thus led the scientific community to include soft computing themes in traditional courses on automatic control, telecommunications and computer science.

At the Engineering Faculty of Catania University, in the various degree courses, and in particular in those of electronic engineering and computer engineering, even from the beginning of the 80s lessons have been given on neural networks, fuzzy control, and genetic optimization. Hence the further need to collect in compact form the most significant themes of soft computing, also for teaching purposes, laying some emphasis both on methodological aspects and applications.

This book aims, therefore, to offer a wide view of the various soft computing features. Moreover the textbook has been written with the aim of being an introduction to the soft computing techniques for educational purposes.

In each chapter, the text is accompanied by a considerable amount of examples. In Chapters 12, 13, 14, and 15 for example, four study cases are outlined; they are considered particularly significant in that they deal with different topics and have different objectives.

The chapter layout is the following:

- Chapter 2 introduces the concept of *fuzzy logic*; in particular, it defines *fuzzy sets* and the *fundamental operators*, *fuzzy algorithms* and some basic techniques for an optimal determination of those algorithms. In addition, an application of fuzzy logic to a simple control problem is reported.
- Chapter 3 is devoted to *fuzzy control*, and in particular to systematic determination techniques for a fuzzy control system and its validation. A study case is also described concerning the fuzzy control of a switching converter, which allows us to examine the proposed techniques in greater depth.
- Chapter 4 contains the basic *artificial neural networks* theory. For some main typologies, e.g., the *multilayer perceptron* (MLP), the non-supervised networks and the *radial basis function* (RBF), the *learning algorithms* are reported in detail. Furthermore, numerical examples are described regarding the interpolation of non-linear maps by multilayer perceptrons, and the RBF classification of defects for diagnosis on-board railroad trains.
- Chapter 5 deals with neural network applications for identifying non-linear dynamic systems and for complex system control. The use of MLP is viewed in greater depth with an example.
- Chapter 6 introduces *evolutionary-type optimization algorithms*, with particular reference to *genetic algorithms*. In this chapter, too, some numerical examples are reported.

- Chapter 7 describes the basic aspects of *cellular neural networks* (CNN) with their structure, the electric model, and basic results regarding their stability. In addition, classical CNN applications for *image processing* are introduced together with a strategy for automatic determination of templates.
- Chapter 8 is devoted to *complex dynamic systems*, and in particular *chaotic* ones. Apart from the basic definitions, the models of chaotic circuits and systems are reported. The possibility of realizing their dynamics by using CNN is also introduced.
- Chapter 9 deals with *neuro-fuzzy networks*, which are derived from the integration between neural networks and fuzzy logic. Some neuro-fuzzy network structures and their respective learning algorithms are described, and a numerical example of non-linear map interpolation is given.
- Chapter 10 reports some novel results concerning the *fuzzy implementation of cellular neural networks*, which prove to be *universal computation* structures. The applications proposed as study cases regard image processing, the revealing of correlations between characters, the solution of reaction-diffusion equations, pattern generation, and the formation and propagation of spiral waves.
- Chapter 11 is devoted to *fuzzy systems optimization*. The study cases considered regard fuzzy control of greenhouse temperatures and the optimal realization of *fuzzy filters*.
- Chapter 12 is entirely addressed to soft computing techniques for solving the problem of modeling the *acoustic pollution* produced by urban traffic. The problem considered is solved by use of and comparison with all the techniques proposed for modeling, and in particular fuzzy logic, neural networks, and neuro-fuzzy networks.
- Chapter 13 contains some results obtained by soft computing techniques in solving the problem of modeling and control of a *citrus fruit-picking robot*.
- Chapter 14 is also devoted to applications of soft computing in industry, with particular reference to modeling and *control rapid thermal process* (RTP) by neural networks.
- Chapter 15 regards the use of neural networks in a new filed of wide interest: the air pollution monitoring and quality prediction in high-density petrochemical area.
- Chapter 16 includes some concluding remarks and one table that summarizes all the proposed examples classifying them both by chapter and application themes.

Each chapter is accompanied by a detailed bibliographical list that offers the reader the significant references made to soft computing by the international scientific community. Among the most recent sources, special mention should be made of the international journal "Soft Computing" published by Springer-Verlag.

2. Fuzzy Logic

2.1 Introduction

Fuzzy logic, the brainchild of Prof. Lotfi Zadeh [1-5], dates back to 1965; it was seen as a technique to counteract the quantitative-type ones that had been successfully utilized till then for analyzing systems where behavior could be described by laws of mechanics, electromagnetism, and thermodynamics. The principle inspiring the theory is known as the principle of incompatibility and states that, as a system becomes increasingly complex, the possibility of obtaining a precise description of it in quantitative terms decreases.

However, in these cases it is often possible to give an imprecise representation of the system's behavior by means of a linguistic description, typical of the human brain, which allows us to synthesize the, at times contrasting, information available, and to extract that needed for the task.

Fuzzy logic is based on the premise that the key elements in the activity of human thinking are not numbers but rather indicators of *fuzzy sets*, *i.e.*, of classes of objects in which the transition between membership and non-membership to the class is a gradual, or even distinct one. In this way, every element belongs to a set with a determinate *degree of membership*.

According to Zadeh, the continual presence of imprecise concepts in thinking suggests the idea that human reasoning is set up on imprecise logic which uses fuzzy sets, connectives, and implications, rather than on binary logic. What is characteristic of fuzzy logic is that it allows imprecise concepts to be dealt with in a well-defined way.

The use of fuzzy logic allows the power, typical of non-linear calculation methods, to be combined with the possibility of representing reality by utilizing a language similar to that of man. In fact, this is a fascinating theory for representing more or less complex phenomena by defining a certain number of fuzzy sets that are elaborated by means of the appropriate connectives. The cause-effect links regulating the process will be described by *fuzzy implications*.

In particular, when solving problems of complex system control, fuzzy logic, through techniques that will be described in this and the next chapter, allows us to determine a control system which combines solutions, proposed by an expert in the system to be controlled, with the power of optimization techniques and the possibility of implementing the control laws obtained on specific hardware [6-10].

2.2 Fuzzy Sets

The basic element of fuzzy logic is the *fuzzy set* that, as the name itself suggests, represents a generalization of the classical concept of a set. Formally, it can be stated that:

*A **"fuzzy"** set A is a collection of objects from the universe of discourse U having some property in common. The set is characterized by a **membership function** μ_A:U → [0,1], which associates with every element y of U a real number $\mu_A(y)$ belonging to the interval [0,1]; this represents the degree of membership of y to the fuzzy set A.*

In general, the greater the value of the function with which the element *y* belongs to the fuzzy set *A*, the greater is the evidence that the object *y* belongs to the category described by the set *A*.

The set of points of *U* for which the membership function to a given fuzzy set *A* is positive is called the *support* of *A*.

Singleton is the definition given to a fuzzy set whose support is only one point of *U*.

The definition of fuzzy set just given allows us to systematically use *linguistic variables*, *i.e.*, variables which assume adjectives, specified by fuzzy sets, as values.

For example, supposing that the linguistic variable *temperature* must be defined, and that one wants to represent the linguistic concept *temperature equal to ca 150 °C*. This concept can be expressed, using traditional logic, by a binary membership function, or, that is to say, one that can assume only the values 0 or 1: if $\mu(T) = 0$, the temperature *T* does not belong to the specified set, e.g., to the interval [135 °C, 165 °C]; if instead it does belong, then $\mu(T) = 1$.

Within the framework of classical sets utilizing a rectangular function, defined in the universe of discourse, that can be described as reported in Figure 2.1.

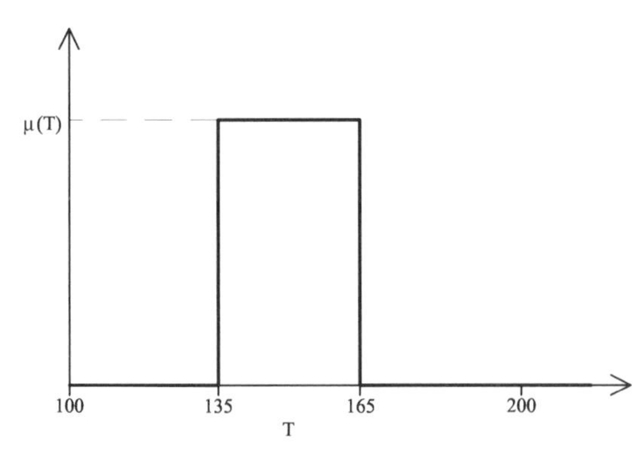

Figure 2.1. Representation of the "*temperature equal to ca 150 °C*" set obtained utilizing the concepts of classical sets

On the other hand, a fuzzy set which represents the same concept can utilize membership function values μ(T) that vary continuously in the interval [0,1], expressing the concept of temperature equal to *ca* 150 °C in a more adequate way than that attributed to such an expression by the human brain; for example, a fuzzy set suitable for representing this concept can be defined as depicted in Figure 2.2.

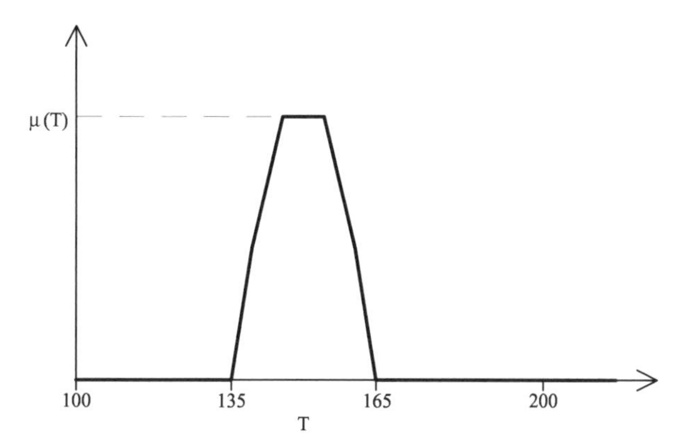

Figure 2.2. Representation of the "*temperature equal to ca 150 °C*" set obtained utilizing the concepts of fuzzy logic

An ordinary set is thus precise in its meaning and presents a definite transition from membership to non-membership.

A fuzzy set, instead, allows us to represent the imprecision of a given concept by means of the degree of the membership function.

Similarly, it is possible to represent by means of fuzzy sets the concept of *high, medium and low temperature*, referred to the universe of discourse determined by the application, and defining the membership functions of the three fuzzy sets as shown in Figure 2.3.

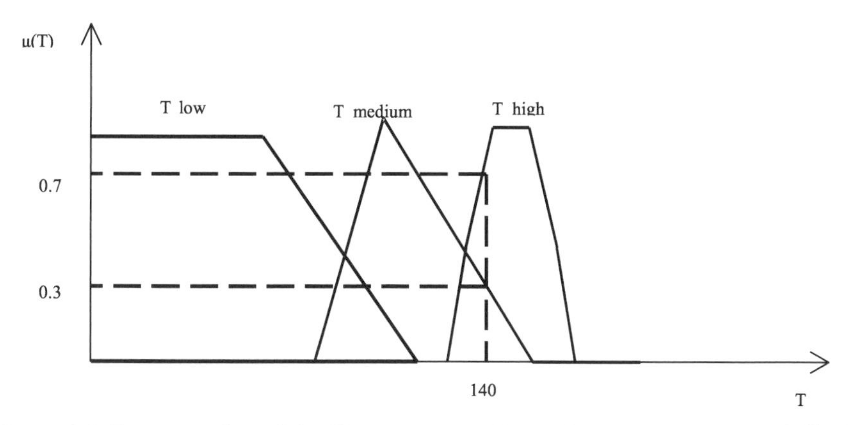

Figure 2.3. Representation of the "*low temperature*", "*medium temperature*", and "*high temperature*" fuzzy sets

As can be seen from the figure, a temperature of T=140 °C has a degree of membership $\mu(HIGH)= 0.7$ to the fuzzy set that represents a *high* temperature, a lesser degree of membership, equal to $\mu(MEDIUM) = 0.3$ to the set defining *medium* temperature, and a zero degree of membership to the set representing the *low temperature*.

The choice of the membership functions for each fuzzy set is wholly subjective and closely connected to the application. In many cases, the choice of the membership function can represent a problem and it is preferred to fall back on the indications of some expert in the field in question, or to optimization techniques [9], [11-12].

2.3 Fuzzy Sets Operations

Within the framework of classical sets, given two sets A and B, the following operations are defined:

$$\begin{aligned} &\text{union: } A \cup B = \{u \in U | \, u \in A \text{ or } u \in B\} \\ &\text{intersection: } A \cap B = \{u \in U | \, u \in A \text{ and } u \in B\} \\ &\text{negation: } \bar{A} = \{u \in U | \, u \notin A\} \end{aligned} \tag{2.1}$$

Analogous operations are defined within the framework of fuzzy sets.

Considering the membership functions of every fuzzy set, the operations mentioned above are defined in the following way:

$$\begin{aligned} &\chi_{A \cup B}(u) = \max(\chi_A(u), \chi_B(u)) = (\chi_A(u) \vee \chi_B(u) \\ &\chi_{A \cap B}(u) = \min(\chi_A(u), \chi_B(u)) = (\chi_A(u) \wedge \chi_B(u) \\ &\chi_{\bar{A}}(u) = 1 - \chi_A(u) \end{aligned} \tag{2.2}$$

where $\chi_A(u)$ and $\chi_B(u)$ are the membership functions of the two fuzzy sets A and B (it should be noted that generally the terms *fuzzy set* and *membership function of the fuzzy set* are used equivalently).

It can be shown that with the definitions adopted, the laws of De Morgan and the properties of absorption and idempotency continue to have formal validity; however, in the case of fuzzy sets we have:

$$\begin{aligned} &A \cup \bar{A} \neq U \\ &A \cap \bar{A} \neq \emptyset \end{aligned} \tag{2.3}$$

That was to be predicted, given that fuzzy theory does not foresee the *dichotomy* characteristic of set theory. Apart from the operators already defined, others can be considered which, however, are used more rarely. The previously defined operations can be made clear with a few examples.

Let us suppose that in the universe of discourse of dimension *x* the two fuzzy sets *x is medium* and *x is large*, reported in Figure 2.4, are defined.

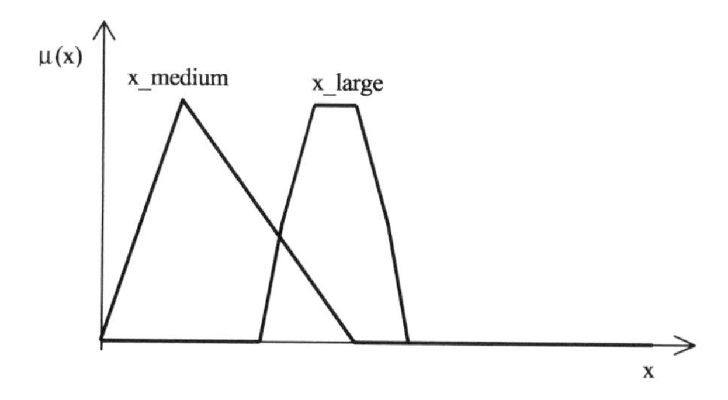

Figure 2.4. Membership function of the two fuzzy sets "*x is medium*" and "*x is large*"

Figure 2.5 reports the membership function of the set obtained from the operation of union "*x is medium or x is large*".

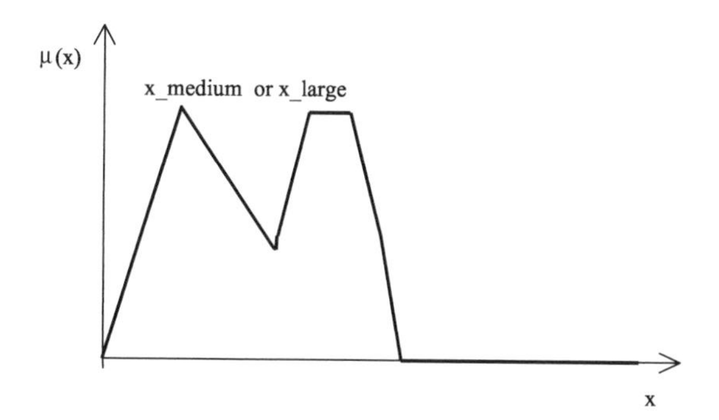

Figure 2.5. Membership function of the fuzzy set "*x is medium or x is large*"

Figure 2.6 reports the membership function of the set obtained from the operation of intersection "*x is medium and x is large*".

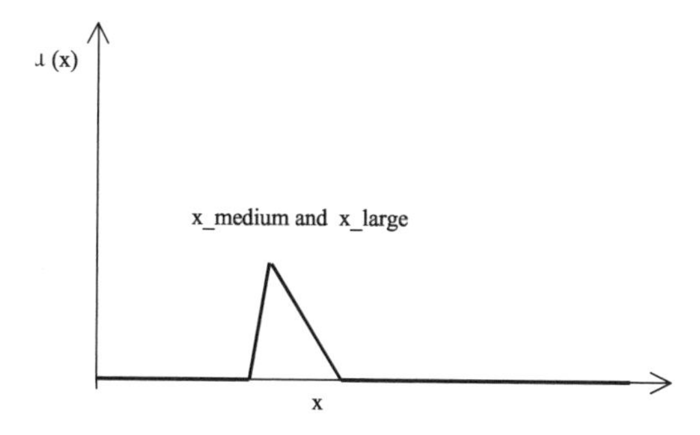

Figure 2.6. Membership function of the fuzzy set "*x is medium and x is large*"

Finally, Figure 2.7 reports the membership function of the set "*x is not large*".

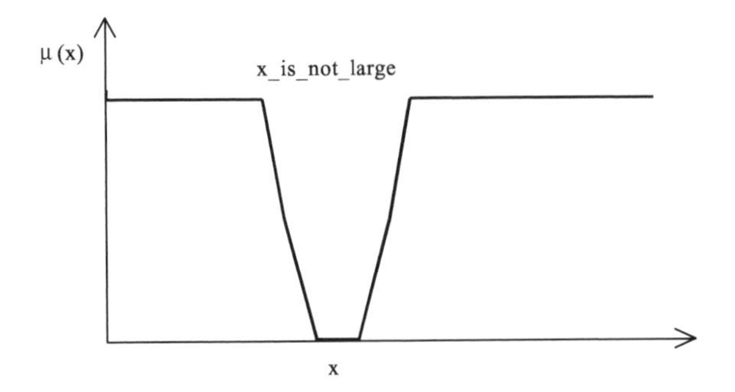

Figure 2.7. Membership function of the fuzzy set "*x is not large*"

2.4 Fuzzy Algorithms

In order to use fuzzy sets for representing the behavior of a system, the concepts of *fuzzy implication* and *fuzzy algorithm* should be introduced.

In its simplest form, a fuzzy implication can be expressed by:

if x **is** A **then** y **is** B;

where both A and B are fuzzy sets. In such an expression, two distinct parts can be recognized; the term:

if x **is** A

is called the *antecedent*, whereas the part:

then y **is** B

is called the *consequent* of the rule. As in classical logic, the result of an implication is governed by the rule of *modus ponens*: *i.e.*, the truth of the implication depends on that of the premise.

Although the logic introduced by Zadeh retains this postulate, it should not be forgotten that the antecedent of the implication is generally characterized by a degree of truth lying in the interval [0,1]. The procedure for calculating the degree of truth, or degree of activation of the antecedent, is known as the *fuzzification* of a numerical value.

By extending the *modus ponens* principle, one admits that a rule, and therefore its consequent, cannot contain more truth than does its antecedent. That is translated by considering as an implication output the fuzzy set obtained by multiplying the fuzzy set B by the degree of activation of the antecedent (the product method) or by cutting B at the height identified by such a value (truncation method).

An example of this is given in Figure 2.8 which reports the rule:

if x is medium then y is large

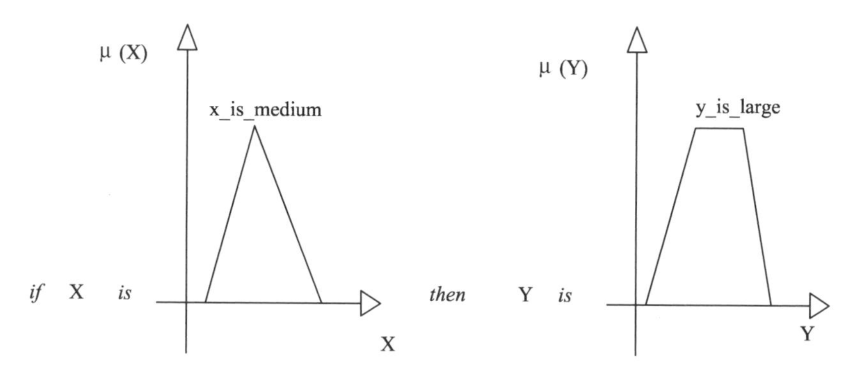

Figure 2.8. Graph of the fuzzy rule *if x is medium then y is large*

Figure 2.9 reports the fuzzy set associated with the implication mentioned above corresponding to a numerical value of magnitude x and obtained by using the two previously defined inference techniques. This fuzzy set is represented by using a bold solid line

The definitions introduced can be extended to fit the case, something very common in practice and particularly so in signal processing, in which the signal chosen to represent the output of a phenomenon depends on several variables; in fact, one might think that in these cases the consequent depends on all the conditions expressed by the antecedent, and thus that these are connected to each other by an *and* fuzzy operation.

In its most general form, the implication assumes the following structure:

if x **is** A **and** y **is** B... **and** n **is** N **then** k **is** K (2.4)

where A, B... N and K are fuzzy sets defined in suitable universes of discourse.

Analogously with what has been previously defined, the term:

if x **is** A **and** y **is** B... **and** n **is** N

is still called antecedent, whereas the construction:

then k **is** K

forms the consequent of the implication.

To determine the degree of activation of the antecedent μ in the case in question, following the *modus ponens* principle it is enough to consider in correspondence with the input values *x, y,… n* the value $\mu = min\ (\mu_A(x), \mu_B(y),\dots \mu_N(n))$.

This value will be used, either with the product method or the truncation method, to determine the output of the rule:

if x is medium and y is large then k is large

reported in the graph of Figure 2.10.

Figure 2.11 shows the output of this rule, corresponding with two numerical values of the magnitudes *x* and *y* and assuming that the truncation method will be used.

In this case, too, the fuzzy set determined is represented by using a bold solid line. Instead, a fuzzy algorithm is formed by a number of *fuzzy rules* which, when evaluated as a whole, allow a given phenomenon to be modeled.

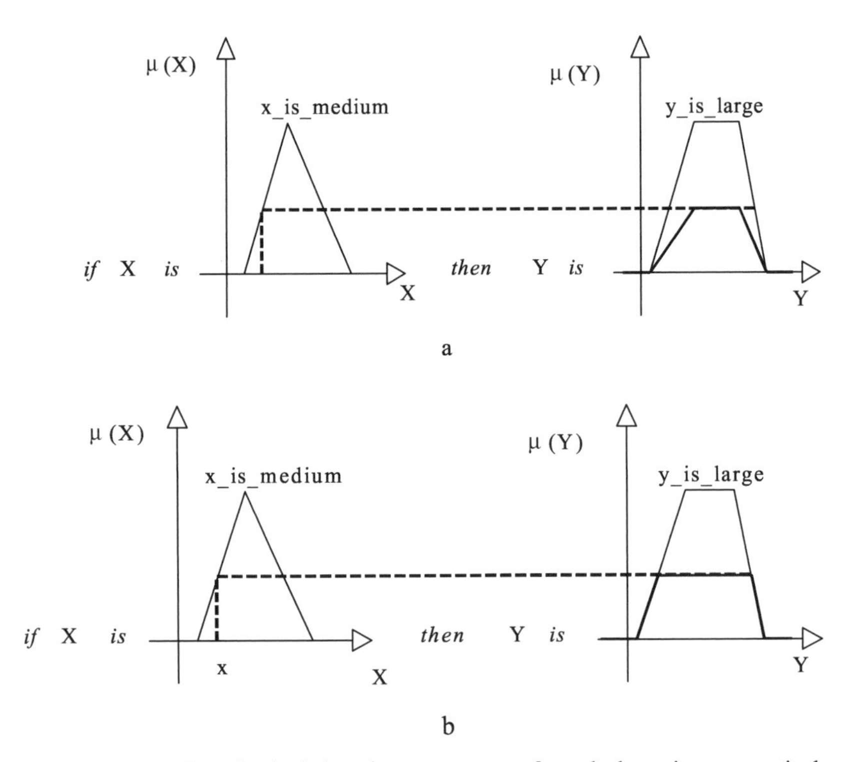

Figure 2.9. Examples of calculating the consequent of a rule by using, respectively: a. the *product method* and b. the *truncation method*

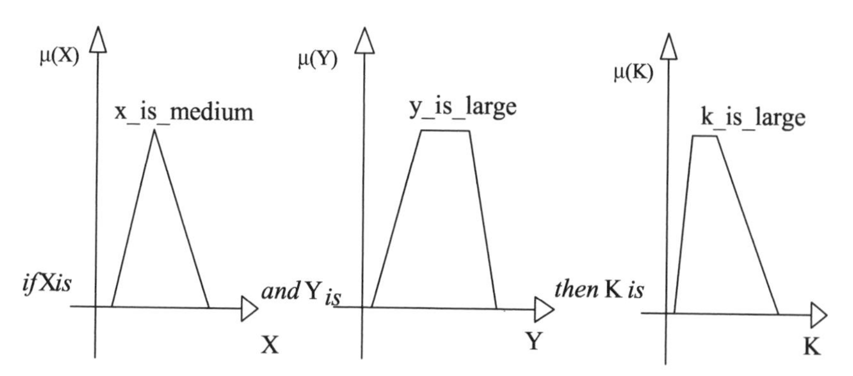

Figure 2.10. Graphic representation of the rule: *if x is medium and y is large then k is large*

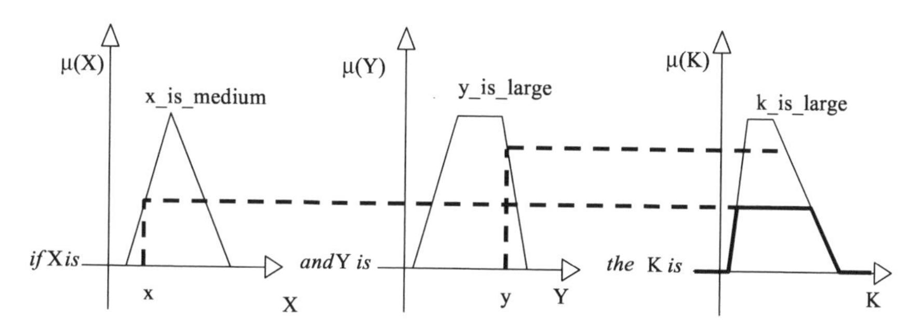

Figure 2.11. Example of the output calculation of a fuzzy implication containing two fuzzy sets in the antecedent corresponding to the input values *x, y*

An algorithm with *m* rules defined on *n* variables then assumes the following general form:

R_1: *if* x_1 *is* A_{11} *and...* x_n *is* A_{1n} *then* y *is* B_1

R_2: *if* x_1 *is* A_{21} *and...* x_n *is* A_{2n} *then* y *is* B_2

....

....

R_m: *if* x_1 *is* A_{m1} *and...* x_n *is* A_{mn} *then* y *is* B_m (2.5)

where the meaning of the symbols is obvious.

The problem now arises of defining how to calculate the output of such an algorithm.

It can be seen that the various rules supply an output value in the form of fuzzy sets, and each one independently of the others. Furthermore, each rule contributes to the building up of certainty regarding the value assumed by the output variable.

Hence it seems logical to state that the output of a fuzzy algorithm be determined by calculating the *or* fuzzy operation of the outputs associated with the individual rules.

Let us consider, for example, the following algorithm made up of two rules:

R_1: *if x is medium and y is large then k is large*

R_2: *if x is large and y is large then k is very large*

and represented in the graph of Figure 2.12.

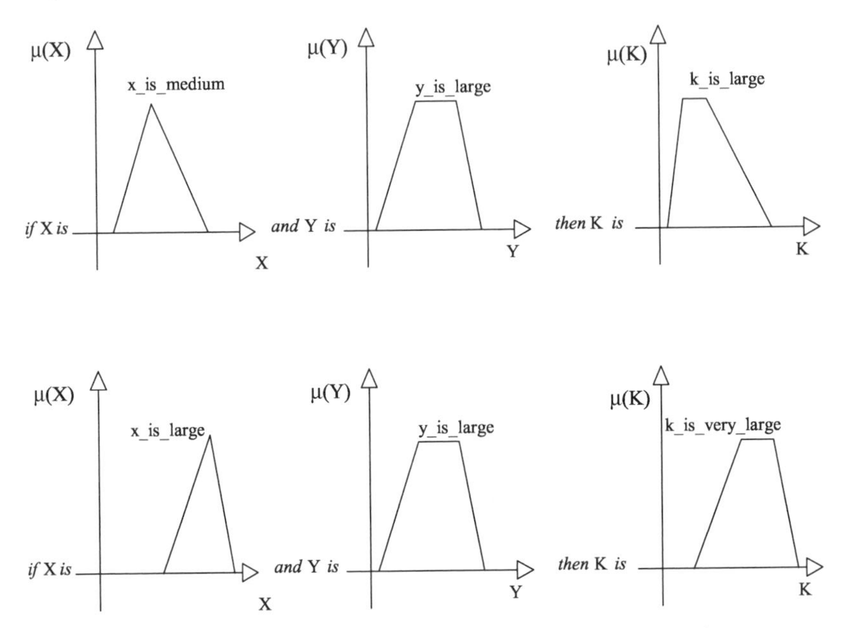

Figure 2.12. Example of a fuzzy algorithm made up of two rules

Assuming that we have measured two values for the magnitudes *x* and *y*, and using the *truncation inference* method, we then obtain, for the output of such an algorithm, a fuzzy set like that reported in the graph of Figure 2.13, where the output of the whole algorithm is represented by the bold solid line represented below on the right.

From what has so far been stated, it is clear that the output of a fuzzy algorithm is made up of a fuzzy set.

Since in practical problems, and in particular when analyzing signals, we describe a magnitude by using a real number, an additional operator should be introduced which associates a real number with the fuzzy set; this number, according to a certain criterion, should be suitable for representing the informational content of the fuzzy set.

This operation is known as *defuzzyfication* and forms the link between the fuzzy world and that of real numbers, with which we have generally been used to describe variables. For this operation, various definitions have been proposed in the literature. In particular, in what follows, the so-called *centroid defuzzyfication* method will be used: according to this method, the fuzzy set is represented by the barycenter of the figure which represents the membership function of the set itself. Hence, we have:

$$k_c = \frac{\sum u(i)\mu(u(i))}{\sum \mu(u(i))} \tag{2.6}$$

with the sum being extended to those elements of the fuzzy set $u(i)$ which have a non zero membership function.

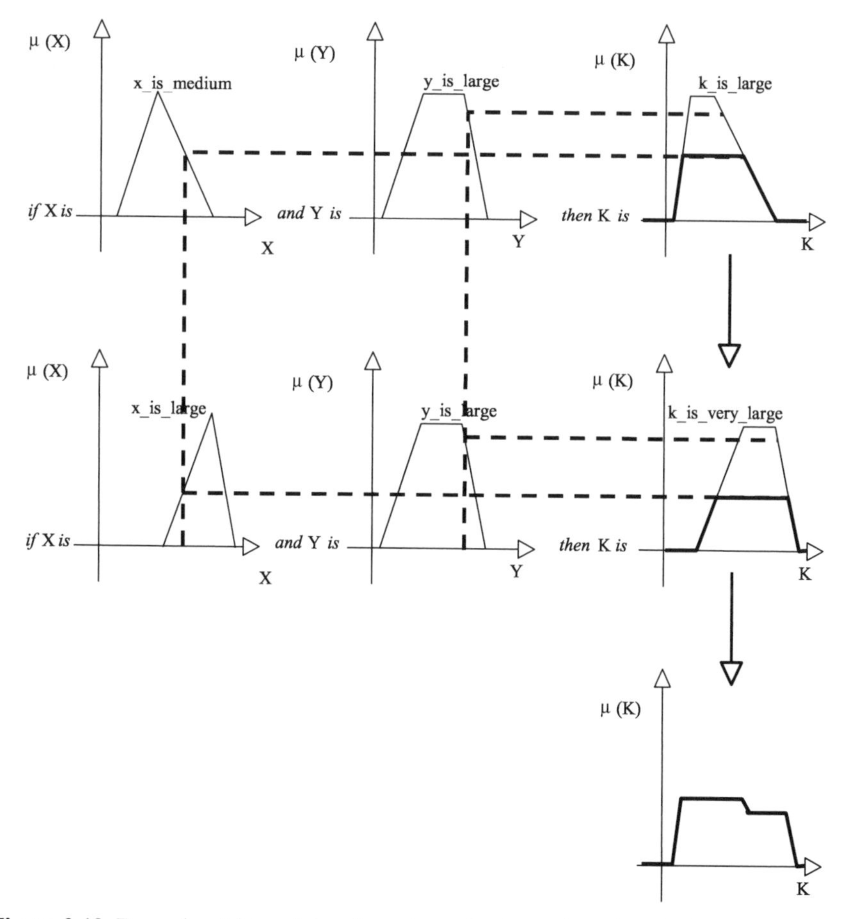

Figure 2.13. Example of determining the output of a fuzzy algorithm obtained by use of the truncation method

2.5 The Fuzzy Algorithm with a Linear Consequent

In fuzzy logic applications, use is frequently made of a structure which is slightly different of the fuzzy implications. In particular, the fuzzy rules considered assume the following form according to what was proposed by Sugeno [11].

$$\textit{if } x_1^i \textit{ is } A_1^i \textit{ and } x_2^i \textit{ is } A_2^i \textit{ and} \ldots \textit{and } x_n^i \textit{ is } A_n^i \textit{ then } y^i = g^i\left(x_1^i, x_2^i, \ldots, x_n^i\right) \tag{2.7}$$

The model introduced presents a structure that can be considered a particular case of the general model introduced by Zadeh. It displays the following peculiarities:

- the membership functions for the fuzzy set of the antecedents are all piece-wise linear functions, neither increasing nor decreasing;
- the consequent is expressed by a real number which can be considered a particular fuzzy set having unitary membership function on a single point of the universe of discourse, *i.e.*, a *singleton.*

In addition, it is presumed that the consequent is obtained as a function of the values assumed by some, or possibly all, the variables occurring in the antecedent.

In the structure proposed in [11], it is assumed in particular that the function $g^i(.)$ is linear.

Thus, the *i*-th implication assumes the form reported in the graph of Figure 2.14.

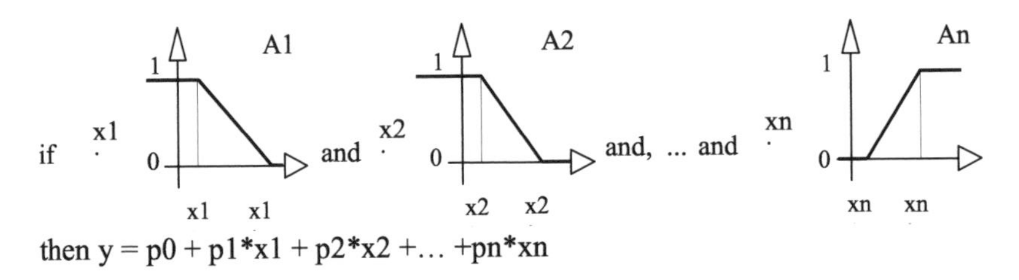

Figure 2.14. Example of the implication proposed in [11]

If the fuzzy algorithm contains *m* rules of the previously introduced form, by applying the centroid *defuzzyfication* method, one finds that the output *y* must be calculated, according to the degree of activation of the individual rules, as a weighted average of the respective y_i outputs:

$$k_c = \frac{\sum_{i=1}^{m} \mu(y_i) y_i}{\sum_{i=1}^{m} \mu_i} \tag{2.8}$$

2.6 Determining the Fuzzy Algorithm

The model just described is of considerable interest in that in [3] a procedure was proposed for determining those parameters that characterize the fuzzy algorithm, in the event that one wishes to determine a set of rules which represent the model of a given system, and that a certain number of input/output measurements of the system to be modeled is available.

With regard to this, it should be noted that it has been demonstrated that a set of fuzzy rules is, with an arbitrary degree of accuracy, able to approximate any non-linear function [13].

In particular, the fuzzy model introduced in the previous section depends on the number and shape of fuzzy sets of the antecedent, as well as on the parameters contained in the corresponding consequents [17].

A more thorough description of the procedure required for determining the fuzzy algorithm parameters is given in [11].

In what follows, every algorithm is identified using a string of the type *M(a,... i,... n)*, where *a,... i,... n* are natural numbers associated with the structure of the model, and in particular where each number indicates how many fuzzy sets are considered for each variable.

In order to obtain the fuzzy model, the following points are necessary:

a. to establish which variables must be considered in the premises. In fact, it may be unnecessary to introduce into the antecedents of the rules all the variables that characterize the phenomenon (some variables may appear only in the consequents);
b. to determine the form of the membership functions for the fuzzy sets characterizing the antecedents of the rules, or else to fix for every set the two values x^i_{j1} and x^i_{j2}, introduced in Figure 2.14;
c. to determine the polynomials associated with the consequents of each rule.

Presuming that points a. and b. have been solved, and thus that the number of rules has been fixed together with their structure and the shape of the membership functions, the solution of Point c. is merely one of determining the coefficients of the polynomials $g^i(.)$ that minimize a suitably fixed index. The mean square deviation E_r, of the measured values of the variable k and the corresponding estimates, appears to be an obvious choice:

$$E_r = \sqrt{\sum \left(k_i - \bar{k}_i\right)^2} \tag{2.9}$$

where:

k_i are the measured output values;

$\bar{k}_i$ are the estimated output values.

In the hypotheses considered, the output is a linear function of the consequent parameters. These parameters can be determined by using of the least squares minimization method [14], [15].

Let us now describe the solution of Point b. The shape of the fuzzy sets used in the various rules depends on the values assumed by the two parameters x^{i}_{j1} and x^{i}_{j2} of Figure 2.14; an optimization procedure must therefore be adopted that will minimize the cost function E_r, with respect to those parameters as well. The problem of optimization introduced is of a non-linear nature, so non-linear programming techniques must be adopted; for example, the Nelder Mead method [16].

The entire search procedure of the *very good* model starts with the solution of the Point a. In other words, to solve the two remaining points, it is necessary to fix how many fuzzy sets are required for each variable.

This problem is of a combinatorial kind, and no general solution exists. In agreement with what was suggested in [11], a heuristic criterion could be used in searching for the structure of the model. The entire search procedure starts with examining the simplest algorithms possible, *i.e.*, models of the type $M(0,\dots 2,\dots 0)$, containing only two fuzzy sets defined on the i-th variable. For each model, then, the consequent parameters and the shape of the membership functions can be calculated, on the basis of what was described above, and then the value of the Er index may be determined. The algorithm affording the minimum value for Er is used for producing more complex algorithms. These are, in fact, generated by this doubling the number of fuzzy sets defined on the generic variable, or by introducing two fuzzy sets for the variables not yet considered.

The procedure described is repeated for as long as the index in Equation 2.9 does not reach a value that is considered satisfactory.

2.7 An Example: Controlling the Inverse Pendulum

Given an inverse pendulum free to move, we must find the torque to apply at the base so that the pendulum is aligned vertically, starting from the initial position $\{x(0), y(0)\}$.

The input variables needed for determining the torque are the angle of the pendulum with respect to the vertical, which we can call "x" (considering as positive the angles starting clockwise from the vertical), and the angular speed which we can call "y".

The output variable is the torque applied at the base "z". Let us define, for example, for each variable five membership functions: *very negative, slightly negative, zero, slightly positive,* and *very positive.*

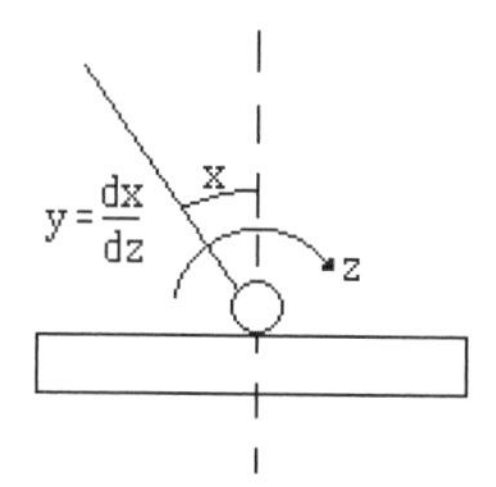

Figure 2.15. Inverse pendulum

The rules can be determined thanks to the knowledge and intuitive powers of an expert. Let us consider, for example, that the pendulum is in the vertical position (the angle is zero) and does not move (the angular speed is zero). Evidently, this is the desired output value, and thus there is no need to perform any control action (the torque must be zero).

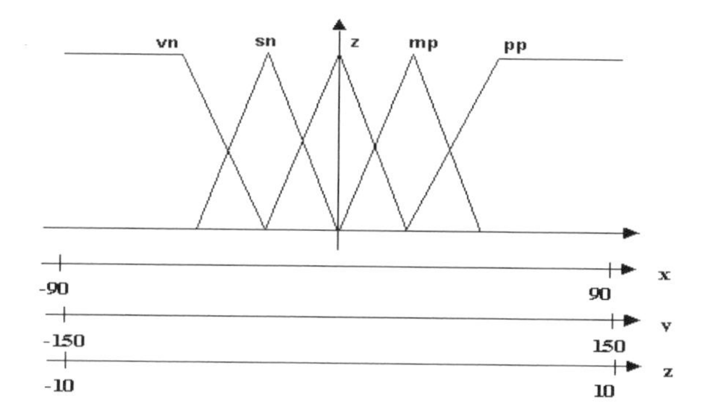

Figure 2.16. Inverse pendulum membership functions

Let us consider another case: the pendulum is in a vertical position as before, but moving at slow speed in the positive direction. Obviously we must compensate the motion of the pendulum by a applying a low-intensity torque in the opposite direction. We have constructed two rules that can be expressed more formally in the following way:

If the angle is zero and the angular speed is zero then the torque is zero

If the angle is zero and the angular speed is slightly positive then the torque is slightly negative

A generic rule is of the type:

If x is A_i and y is B_j then $z = C_k$

with $i, j, k = 1 \ldots 5$.

By effecting all the combinations with the antecedents, 25 rules of this type could be defined, but the knowledge of the problem leads us to identify only 11 of them, those given in Table 2.1, where *very negative* is abbreviated with **vn,** *slightly negative* with **sn**, *etc.*

Even if in the proposed example, the rules have been determined starting from the knowledge of an expert of the system, it is always best to follow up the phase of determining the fuzzy sets with an optimization phase which will improve the performances. With this in view, the evolutionary optimization algorithms introduced in Chapter 6 are often used; these allow the shape and position of the membership functions to be codified very simply, and the optimal values to be determined with respect to a suitable index representing the goodness of the algorithm.

Table 2.1. Rules used for controlling the inverse pendulum

y / *x*	vn	sn	*z*	pp	Mp
Vn			**mp**		
pn			**pp**	*z*	
z	**mp**	**pp**	*z*	**sn**	**vn**
pp		*z*	**sn**		
mp			**vn**		

2.8 References

1. Yager RR, Zadeh LA editor. An Introduction to Fuzzy logic Applications in Intelligent Systems. Kluwer Academic 1992
2. Bezdek JC. Analysis of Fuzzy Informations. CRC Press 1987
3. Zadeh LA. The Concept of a Linguistic Variable and its Application to Approximate Reasoning. Part I, Inf. Sci. 1975; 8: 199-249
4. Zadeh LA. The Concept of a Linguistic Variable and its Application to Approximate Reasoning. Part II, Inf. Sci. 1975; 8: 301-357
5. Zadeh LA. The Concept of a Linguistic Variable and its Application to Approximate Reasoning. Part III, Inf. Sci. 1976; 9: 43-80
6. Tong RM. A Control Engineering Review of Fuzzy System. Automatica 1977; 13: 559-569
7. Rizzotto G, Lavorgna M, Lo Presti M. Metodologie per la Sintesi e l'Analisi di Controllori Fuzzy. Cavallotto Edizioni, 1996
8. Jamshidi M, Vadiee N, Ross TJ. Fuzzy Logic and Control: Software and Hardware Applications. Prentice Hall, Englewood Cliffs, NJ, 1993; 2
9. Gupta MM. Fuzzy Neural Network Approach to Control Systems. Proc. Am. Control Conf. San Diego, 1990; 3: 3019-3022
10. Sugeno M. Industrial Applications of Fuzzy Control. North-Holland, Amsterdam, 1985

11. Takagi T, Sugeno M. Fuzzy Identification of Systems and Its Applications to Modeling and Control, IEEE Trans. on Systems, Man and Cybern. 1985; 15: 1
12. Jang JSR, Sun CT, Mizutani E. Neuro-Fuzzy and Soft Computing. Matlab Curriculum Series
13. Kosko B. Fuzzy Systems as Universal Approximators. Proc. IEEE Int. Conf. Fuzzy Syst. San Diego, 1992; 1153-1162
14. Hsia TC. System Identification: Least-Squares Methods. D. C. Heath and Company, 1977
15. Ljung L. System Identification: Theory for the User. Prentice Hall, Upper Saddle River, NJ, 1987
16. Press WH, Flannery BP, Teukolsky S A, Vetterling W T. Numerical Recipes, The Art of Scientific Computing. Cambridge University Press, Cambridge, 1986; 289-293
17. Baglio S, Fortuna L, Graziani S, Muscato G. Membership Function Shape and the Dynamic Behavior of Fuzzy System. International Journal of Adaptive Control and Signal Processing. 1994; 8: 369-377

3. Fuzzy Control

3.1 Introduction

The strategy of fuzzy control proposed by Zadeh in the early 60s rapidly became an object of growing interest in the field of *control systems*. The utility of this kind of control technique was thus explained by Zadeh.

> *One of the reasons why human beings have a better control capacity than do present-day machines is that human beings are able to make decisions on the basis of imprecise linguistic information.* (L. Zadeh, 1965).

In fact, one of the limitations of the *ordinary control systems*, and especially those more widespread on an industrial scale, such as the so-called *standard PID controller*, is precisely their *lack of flexibility* both with regard to the process they have to control and to their possibility of integrating a formal type of knowledge (that used for synthesizing the controller) with other heuristic-type knowledge that the experts may derive from their practical knowledge of the plant. If one wishes to compare *fuzzy control* with other traditional control techniques, comparable in simplicity and low cost, and used industrially, e.g., the PID technique, fuzzy control proves more *robust* in the majority of cases and gives higher performance in presence of variations in parameters, load, and external disturbance [8-9]. These characteristics will be evidenced by many examples described in the following chapters.

Fuzzy control is particularly versatile in all cases where the knowledge of an expert can be called upon. In many industrial processes, supervision of the human operator is of fundamental importance, but paradoxically, the interest in fuzzy control has not been expressed by experts in this field, most of whom seem to have displayed a fair amount of skepticism for the need of such a control technique in a scenario already particularly rich in methodologies; instead, the interest stems from market applications. In fact, many companies have produced fuzzy control systems in numerous sectors ranging from the automobile market (injection control, ABS, intelligent suspension, exhaust gas control, anti-theft devices, *etc.*) to those of household appliances (washing machines, refrigerators, vacuum cleaners, air conditioners, *etc.*), HI-FI (environment effects, filtering, self-focusing, *etc.*), industrial production systems (quality control, diagnostics, vibration control, power systems, robotics), and to biomedical equipment and telecommunication (receivers, management of nets, traffic supervision, compression of information). In fact, the attitude of many researchers can be understood if one considers that the theoretical apparatus for fuzzy control is rather *atypical* compared with the usual forms, which are mostly in the function of developing applications, and displays considerable shortcomings regarding problems of analysis, such as that of stability, which have always been considered indispensable for the correct development of systems control. On the other hand, it should be recognized that it is precisely this atypical feature of fuzzy control, based on recently introduced concepts, that has required a considerably great effort, in many aspects a still on-going one. In this chapter, in the interests of brevity, we will, avoid embarking upon questions of principle or those of a strictly technical nature, since our main aim is to illustrate the basic concepts for synthesizing a fuzzy control system.

3.2 A Systematic Approach for the Design of a Fuzzy Control System

Fuzzy controller *synthesis* consists in formalizing and characterizing rules for describing the controller. In the absence of any *systematic methodology* for modeling and designing, the problem of extracting such rules requires the presence of an *expert* who, on the basis of *heuristic rules* inferred from his experience, will fix the number of fuzzy sets, the shape of the relative membership functions, and above all describe the control algorithm.

In order to improve the performance of the controlled system, considerable effort is needed to refine the fuzzy rules. Since no analytical relationship exists between the desired performance and the fuzzy controller structure, the calibration of the fuzzy controller can be effected either on the basis of a *trial and error* type of heuristic approach, or on one of *numerical optimization*.

Starting from the consideration that fuzzy systems are *non-linear* ones, the classical techniques valid for linear systems generally cannot be applied; thus, new methodologies need to be developed, aimed at the solution and simplification of fuzzy controller synthesis.

In this chapter, two procedures are put forward for automatic synthesis of a fuzzy controller, and for analyzing its stability characteristics and closed-loop performance. The flow chart shown in Figure 3.1 schematizes the complete approach proposed for the fuzzy controller project.

The following phases can be identified:

1. *Determining the controller system model*

   The techniques of synthesis and analysis proposed do not impose any limitations on the type of model adopted for describing the system to be controlled. This can thus be of the following types:

   - *mathematical*, if there is complete knowledge of the system's structure;
   - *neural*, if the input-output data pairs are completely characterized;
   - *linguistic*, if there is an input-output data set or an expert to describe the system.

In the absence of a mathematical model, it is therefore possible to build a neural model on a *multilayer perceptron* net or a *fuzzy model*. As described in the previous chapter, fuzzy modeling techniques are based on classical (*least squares*) or neural (*neuro-fuzzy nets*) identification models applied derived from a set of measurements made on the process.

2. *Determining the control laws*
   In this phase, the control laws are extracted in the form of a *discrete map*. A technique, described in detail in the next sections, is used which allows to map from the input space variables to the controller output. This map is extracted by using *numerical optimization* algorithms. The performances required for the closed-loop system are summarized in a *performance functional* to be optimized for all cells into which the space of the input variables is subdivided.
3. *Determining the fuzzy controller*
   The knowledge base afforded by the discrete control map is translated into a fuzzy rules system which constitutes the controller. For this purpose, several techniques can be utilized:
   - the *least squares* method;
   - *neuro-fuzzy* modeling techniques.
4. *Validating the fuzzy controller*
   In the absence of a mathematical model of the closed-loop system, and in general for non-linear systems, the *stability analysis* methods are of a numerical type.

The fuzzy controller is validated using a technique for identifying non-linear systems, known as *cell-to-cell mapping*, which allows the stability of the controlled system to be studied and the identification of, in particular, *periodic movement*, *bifurcation*, and *attraction domains* in the spaces of the closed-loop system phases. Although other stability criteria have been proposed in the literature, we will here merely outline this method, which has general validity and places no constraints on the structure of the fuzzy controller adopted.

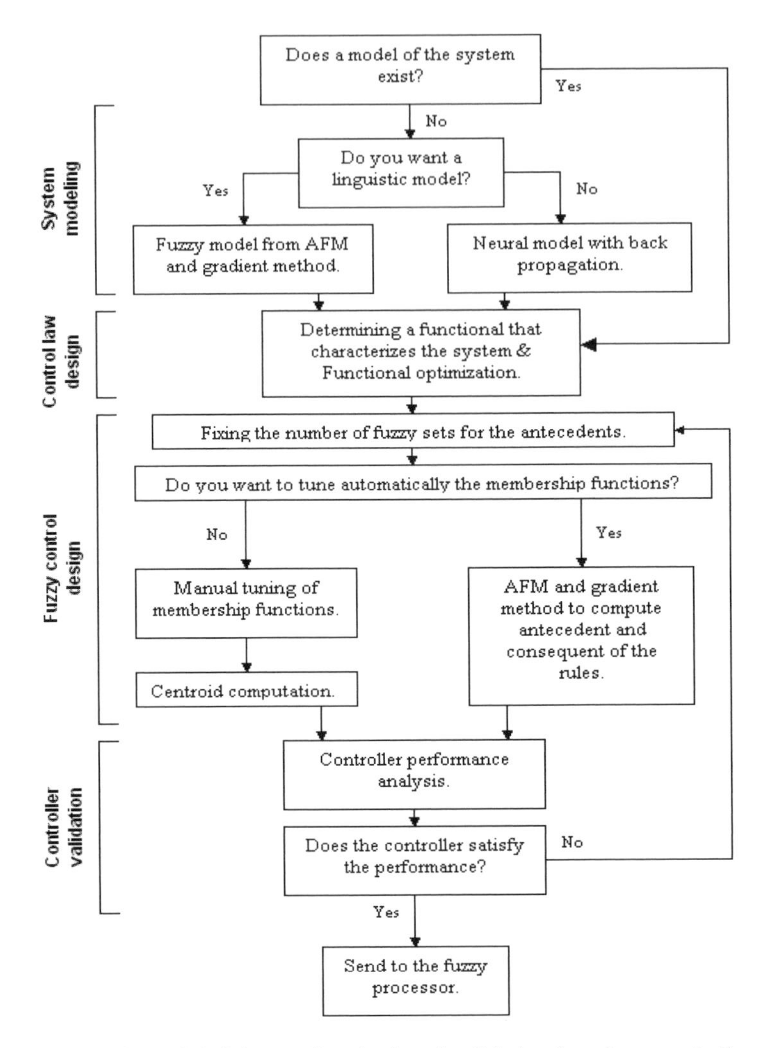

Figure 3.1. Scheme of synthesis and validation for a fuzzy controller

3.3 Synthesis and Validation of a Fuzzy Controller

As already mentioned, in the design methodology for a fuzzy controller and, generally, for a fuzzy system with imposed dynamics, two phases can be identified: i) that of *extraction* of the *control map*, defined in the controller input variable space, giving values in the controller output variable; and ii) *identification* of the *fuzzy model* which implements those laws. In general, this is understood as the fuzzy controller inserted in a *feedback control*, but the techniques described are also valid for a *feed-forward* type of control.

3.3.1 Determining the Control Laws

The first phase, often the formalizing of a human operator's experience, is fundamental for executing the performances required of the system. Generally, the techniques for extracting control laws are based on numerical optimization algorithms. The control map is discrete and is calculated by operating in the *discrete space* of the variables contained in the premises, according to a technique of discreteness through which the space becomes a *continuum of cells* rather than points, and the behavior of every point belonging to a cell is assimilated in that of the cell itself. The approximation and refinement of the control are therefore linked to the discreteness. Evidently, the *upper limit* of the number of cells is connected to the precision of the measurements, to the computational load deriving from it and to the interpolative capacity of the fuzzy model chosen for implementing the controller. The performance of the closed-loop system is markedly affected by the resolution of the discrete map, the interpolative capacity of the fuzzy structure and the optimization procedure. In this chapter, optimization is effected by the well-known *gradient method*, whose algorithm is summarized. The procedure can also be applied to processes for which no mathematical model, but instead a linguistic- or neural-type model exists. Let us suppose, however, that we have a traditional mathematical model of the system to be controlled, in the discrete form:

$$x(k+1) = f(x(k), u, k) \tag{3.1}$$

where $x \in \Re^n$ is the system's state vector, u the input signal, and k the discrete time. The steps to be taken are summarized in the following points:

- the state space is divided into an appropriate number of cells, each of which represents an initial condition of the system to be controlled;
- it is presumed that the behavior of the system is the same for all the initial conditions belonging to the cell and that the points are all identified with one of them, for example the center;
- a functional is chosen for optimization. This functional summarizes the desired performances, such as regime errors, overshoot, *etc.*, and can be given, for example, by the weighted sum of the square errors of the variables to be controlled with respect to the reference values. For each cell, the control variable value $u_o(\cdot)$ of the system which optimizes the functional chosen is calculated.

A simple optimization algorithm is the *gradient descent* that carries out the search for the *minimum* or *maximum* of the functional which resumes the performances, based on a calculation of the *derivative*. In reality, when the functional depends on more than one state variable, the search for the minimum becomes more complex and the use of *multi-objective* or *conditioned optimization* is necessary.

Once the functional $F(\cdot,u)$ has been chosen, the *gradient descent* method can be resumed as follows:

1. the initial value of F_{in} of the functional is calculated for each cell under the action of an initial value of the control variable u_{nom}; if F is given by the sum of the square errors, we obtain:

$$F_{in} = \|X(k+1) - rif\| \tag{3.2}$$

with $X(k+1) = f(X(k), u_{nom})$

where the symbol $\|\ \|$ stands for the norm of a vector;

2. the value u_0 which will minimize the functional is searched for, proceeding with an iterative process according to the following scheme:

2.1. the percentage *perc* is chosen with which to vary u_{nom} in searching for the minimum of F;
2.2. u_{nom} is increased by the percentage chosen:

$$u_0 = u_{nom} + \text{sign } u_{nom} \ perc \tag{3.3}$$

where sign assumes a positive or negative value according to whether it is desired to increase or decrease u_0;

2.3. the system to be controlled is integrated starting from the initial conditions represented by the cell in question under the action of u_0:

$$X(k+1) = f(X(k), u_0, k) \tag{3.4}$$

and the value assumed by the functional F is measured again. If the variation ΔF of F is positive, the sign of the variation of u_0 is changed:

$$\text{sign} = - \text{ sign} \tag{3.5}$$

$$perc = 0.5 \ perc \tag{3.6}$$

$$u_0 = u_0 + \text{sign } perc \ u_0 \tag{3.7}$$

If, instead, ΔF is negative, *sign* does not vary, the speed of variation of u_0, *perc* decreases, and u_0 is calculated again as:

$$u_0 = u_0 + perc \ u_0 \tag{3.8}$$

The procedure continues until the value of the functional reaches the minimum, within the *limits of tolerance* established, or until ΔF becomes less than an *imposed value*, below which it becomes irrelevant to proceed. At the end of the optimization procedure, for each cell we have the optimal value of the control variable u_0 to impose on the system input in order to obtain the desired behavior in one sampling step.

3.3.2 Determining the Fuzzy Controller

Once the procedure of extracting the control law has been completed, this law must be expressed as a *fuzzy algorithm*. Therefore all the parameters characterizing the fuzzy controller must be determined, *i.e.*:

- the *number of fuzzy* sets and the shape of respective membership functions for each variable of the premise;
- the *number of rules*;
- the *shape of the consequents* of each rule.

The most common techniques in the literature for modeling fuzzy systems are generally based on classical methods of *parameter identification* or on *identification techniques* using neural networks. Below, a least squares identification method, already introduced in the previous chapter with regard to Sugeno's model, is described; the techniques based on the use of neural networks will be described in Chapter 9.

For the sake of simplicity, the *number* and *shape* of the fuzzy sets are established *a priori* for each of the controller input variables. The number of all the possible combinations then gives the maximum number of rules:

$$r = n_1 \; n_2 \; \cdots \; n_n \tag{3.9}$$

where n_i represents the number of fuzzy sets defined by the i-th variable. The *shape* of the consequents for each rule still remains to be determined.

It can be seen that once the shape of the fuzzy set membership functions of the antecedents has been fixed, the *degree of activation* for each rule is then determined for each cell. Figure 3.2 reports the example of a second-order system with two fuzzy sets for each state variable. If the *height method* is adopted as that of defuzzification, the *defuzzified output* has the value:

$$y = \frac{\sum_{1}^{r} \mu(R_i) y_{0i}}{\sum_{1}^{r} \mu(R_i)} \tag{3.10}$$

where, with reference to the Sugeno model, we have:

$$y_{oi} = c_{oi} + c_{1i}x_1 + c_{2i}x_2 + \ldots + c_{ni}x_n \tag{3.11}$$

and where $\mu(R_i)$ is the degree of activation of the i-th rule. Equation 3.10 can thus be rewritten as follows:

$$y = \sum_{1}^{r} \alpha_i \left(c_{oi} + c_{1i}x_1 + c_{2i}x_2 + \ldots + c_{ni}x_n \right) \tag{3.12}$$

with

$$\alpha_i = \frac{\mu(R_i)}{\sum_1^r \mu(R_k)} \tag{3.13}$$

where the only unknown factors are the y_{0i}. The value of y coincides with the value u_0 calculated for the cell in question. Writing Equation 3.12 for each of the m cells, we obtain a system of m equations in $(n+1)r$ unknown factors of the type:

$$A\,y_0 = y \tag{3.14}$$

where:

$$A = \begin{bmatrix} \alpha_{11}, & \dots & \alpha_{n1}, & x_{11}\alpha_{11}, & \dots & x_{11}\alpha_{n1}, & \dots & x_{n1}\alpha_{r1} \\ \dots & & & & \dots & & & \dots \\ \alpha_{1m}, & \dots & \alpha_{nm} & & \dots & & \dots & x_{nm}\alpha_{rm} \end{bmatrix} \tag{3.15}$$

$$y_0 = [c_{01}\ \dots,\ c_{0n}\ ,\ \dots\ ,\ c_{0r}\ ,\dots,\ c_{nr}] \tag{3.16}$$

and y is the vector of the control values obtained by the optimization process. The number of lines of A is equal to the number of cells. The solution y_0 of the system (3.14) is obtained by use of the least squares method:

$$y_0 = (A^T A)^{-1}\, A^T\, y \tag{3.17}$$

3.3.3 Validating the Fuzzy Controller

One of the main reasons why fuzzy control was not initially accepted is connected with the absence of any criteria for validating its *stability*. A fuzzy controller is a non-linear system, and thus the traditional methods of analysis available for the class of linear systems cannot be applied to systems controlled by *fuzzy* strategies.

Recently several techniques have been proposed for verifying fuzzy system stability. Generally they impose limitations on the structure of the model, although they have proved to be efficacious. In this section, a methodology, known as cell-to-cell mapping, is described for non-linear systems; it is generally valid and does not place limitations on the structure of the system to be analyzed precisely because it is *numerical*. It should be borne in mind that the closed-loop system is in general a *hybrid* one from the standpoint of modeling: the model of the system to be controlled is not necessarily available in fuzzy form, but may be a classical mathematical, neural model, *etc.*

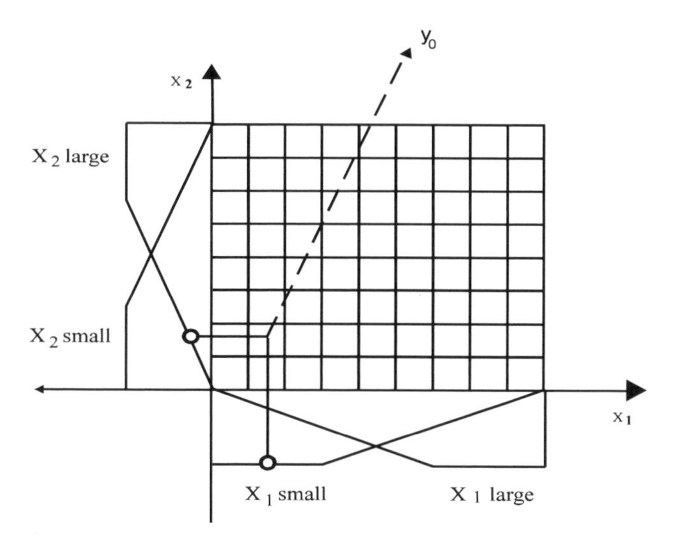

Figure 3.2. Example of calculating "*if x is small and x_2 is small*" on the cell

The case may even occur that no model for the process exists, but only a set of *measurements*. In such cases, the techniques for stability analysis proposed in the literature for a fuzzy system can no longer be applied.

As a numerical identification technique, cell-to-cell mapping proves to be more versatile and is described in this section.

It is known that a complete analysis of non-linear systems is only possible by means of *simulation*. However, a simulation algorithm needs an infinite time to determine the behavior of the system in all the domains of interest.

Cell-to-cell mapping represents a technique for supplying in a reasonable amount of time an *approximate* but *complete* characterization of the system in the state space region it is intended to examine, starting from any model whatsoever, whether it be analytical, linguistic, or formed by some parts described linguistically and by others described analytically. It represents a general strategy for analyzing systems and can therefore be applied to any system [10].

Unlike *point-to-point* techniques, such as *Poincaré maps*, that allow a system to be evaluated in a strictly local environment, the technique described here carries out an analysis on a *finite number of cells*, in which the state space is discrete. If one considers that the imprecision inherent in a knowledge of the model and in the methods of measuring often makes a point-to-point approach superfluous, in the sense that the solution claimed is greater than the real one, there follows the impossibility of considering state variables as being really continuous. Hence the revaluation of analytical techniques, like cell-to-cell mapping, based on making the state space a discrete one.

Therefore, on moving from the continuum to the discrete, each variable is considered a collection of cells rather than a continuity of points, and every point belonging to the cell is associated with the behavior of the same cell to which it belongs. We thus obtain an approximated state space made up of a finite number of cells of appropriate size, where inside each cell the various points lose their own individuality and become confused with a single point that characterizes that cell.

Let us suppose that we have a model in the form:

$$\dot{x}(t) = f(t, x(t)) \tag{3.18}$$

where x is the vector of the state variables and N the system order. Dividing each coordinate axis into a certain number of intervals of size h_i and indicating the interval with the whole of Z_i, this characterizes all the x_i satisfying the relation:

$$(Z_i - \frac{1}{2})h_i \leq x_i \leq (Z_i + \frac{1}{2})h_i \tag{3.19}$$

Given that $Z = (Z_1, ..., Z_N)$ the *coordinate vector* of the cell, a point x belongs to the same cell if x_i belongs to Z_i, for all cases of i. The state space thus becomes a set of cells and the map f is transformed into a map C, which is called *cell mapping*:

$$Z(n+1) = C(Z(n)) \tag{3.20}$$

Identifying, therefore, the generic cell with the points of a Euclidean space having coordinated Cartesian axes with integer values, we can then write:

$$Z = \sum_{i}^{N} Z_i e_i \tag{3.21}$$

where e_i is a unitary vector constituted by zeros and by one in position i *{0,0, ..., 1,0, ..., 0}*.

On the basis of this description of the system dynamics, we then have the following definitions for *equilibrium cell* and *periodic motion*.

A cell Z^* is said to be in equilibrium if we have:

$$Z^* = C(Z^*) \tag{3.22}$$

A sequence of cells $Z^*(j)$ $j = 1, ..., K$ forms a *periodic motion*, *P-K*, of the *period K* if the following relations are satisfied:

$$Z^*(m+1) = C^m (Z^*(1)), \qquad m = 1,2,..., K\text{-}1 \tag{3.23}$$

$$Z^*(1) = C^K(Z^*(1)) \tag{3.24}$$

An equilibrium cell is therefore a periodic motion of period 1.

Given the following definitions of a cell Z' *contiguous* to Z in direction Z_i:

$$Z' = Z \pm e_i \tag{3.25}$$

with a *forward* increment vector of the mapping of C on Z:

$$\Delta_k C(Z) = C(Z + e_k) - C(Z) \tag{3.26}$$

and a *backward* increment vector:

$$\nabla_k C(Z) = C(Z) - C(Z - e_k) \tag{3.27}$$

the *nucleus of equilibrium cells* can then be defined as the most extensive set of cells contiguous to the equilibrium. The size of the nucleus is given by the number of cells forming it. Given a Z^* equilibrium cell, it is isolated if none of the contiguous cells is in equilibrium, that is to say if the following conditions apply:

$$\Delta_j C(Z^*) \neq e \quad \nabla_j C(Z^*) \neq e_j \qquad \text{for each } j=1,2,...,N \tag{3.28}$$

vice versa, if the *forward* increase vector of the equilibrium cell Z^* in direction j is e_j, then the cell contiguous to the equilibrium one in direction j is also an equilibrium one, and *vice versa*. What has been said follows readily from the consideration that since:

$$\Delta_j C(Z^*) = C(Z^* + e_j) - C(Z^*) \qquad\qquad C(Z^*) = Z^* \tag{3.29}$$

then

$$\Delta_j C(Z^*) = e_j \tag{3.30}$$

The case of the *backward* increase vector is analogous.

To estimate the distance between two distinct periodic motions, the *forward* and *backward* increment vectors are defined for the mapping C^m:

$$\Delta_j C^m(Z) = C^m(Z + e_j) - C^m(Z) \tag{3.31}$$

$$\nabla_j C^m(Z) = C^m(Z) - C^m(Z - e_j) \tag{3.32}$$

Let $Z^*(j)$, $j=1,2,...L_1$ and $Z^{**}(j)$, $j=1,2,...,L_2$ be two generic cells making up part of the periodic motions $P\text{-}L_1$ and $P\text{-}L_2$, respectively:

$$C^{L_1+m}(Z^*(1)) = C^m(Z^*(1)) \tag{3.33}$$

$$C^{L_2+m}(Z^{**}(1)) = C^m(Z^{**}(1)) \tag{3.34}$$

Let L be the lowest common multiple between L_1 and L_2, and let us suppose that $Z^*(1)$ and $Z^{**}(1)$ are contiguous. We will then have:

$$Z^{**}(1) - Z^*(1) = e_k \tag{3.35}$$

$$\begin{cases} Z^{**}(j) - Z^{*}(j) = C^{j\text{-}1}\left(Z^{*}(1) + e_k\right) - C^{j\text{-}1}\left(Z^{*}(1)\right) = \\ \qquad\qquad = \Delta_k C^{j\text{-}1}\left(Z^{*}(1)\right) \\ j = 1,2,\dots,L \end{cases} \tag{3.36}$$

$$Z^{**}(1+L) - Z^{*}(1+L) = \Delta_k\, C^L(Z^{*}(1)) = e_k \tag{3.37}$$

A periodic cell *P-K* can then be called *isolated* when, among the cells contiguous with it, none belong to the same periodic motion *P-K*. Formally:

$$\Delta_k C^K (Z^{*}) \neq e_k \qquad \nabla_k C^K (Z^{*}) \neq e_k \tag{3.38}$$

A *P-K* cell is said to be *an isolated periodic cell* if none of the cells contiguous to it belong to any periodic motion:

$$\Delta_k C^{pK} (Z^{*}) \neq e_k \qquad \nabla_k C^{pK} (Z^{*}) \neq e_k \tag{3.39}$$

A *nucleus* of *P-K* cells is the definition given to the set of all the contiguous *P-K* cells, while a *nucleus of periodic cells* is that given to a set of contiguous cells each of which belongs to a periodic motion, even if the periodicity is different.

In analyzing a non-linear system, it is important to estimate the local behavior of the *P-K* cells. Once the norm has been defined as:

$$\|Z\| = \Sigma_i |Z_i| \tag{3.40}$$

the behavior of a system around the equilibrium point Z^* can be predicted. By carrying out an analysis in the generic direction j, the following behaviors can be predicted for the system:

forward repulsive in direction j if: $\left\|\Delta_j C^L(Z^{*})\right\| > 1$

forward neutral in direction j if: $\left\|\Delta_j C^L(Z^{*})\right\| = 1$

forward attracting in direction j if: $\left\|\Delta_j C^L(Z^{*})\right\| = 0$

backward repulsive in direction j if: $\left\|\nabla_j C^L(Z^{*})\right\| > 1$

backward neutral in direction j if: $\left\|\nabla_j C^L(Z^{*})\right\| = 1$

backward attracting in direction j if: $\left\|\nabla_j C^L(Z^{*})\right\| = 0$

Finally, the *attraction domain* at r steps of a *P-K* motion is defined as the set of cells which lies at r steps from the motion, and its *domain of global attraction* as being constituted by its domain at r steps with $r \to \infty$.

Finally, since the number of cells in the domain of analysis chosen is *finite*, it can be verified that each of them belongs to a periodicity of motion K, or that it lies r steps from a periodic motion, or that it diverges in a finite number of steps on leaving the domain in question.

In the last case, the behavior of the system associated with the cell can be considered unstable in the domain in question.

The analysis carried out thus allows us to identify the behavior associated with every cell by means of:

- the *group number* $G(Z)$ that identifies the periodic motion to which the cell belongs and the domain of attraction associated with it;
- the *step number* $S(Z)$ that indicates the number of integration steps separating the cell in question from the periodic motion of the group to which it belongs;
- the *periodicity* $P(G(Z))$ that indicates the number of cells making up the periodic motion of the group G.

Thus, once the state variables of the closed-loop system controlled by the fuzzy regulator have been identified and once the variation ranges for each of them have been fixed, the *stability regions* of the domain in question can be characterized and the equilibrium points and periodic cycles be identified.

3.4 An Example: The Control of a Switching Converter

In order to demonstrate the applicability of fuzzy control logic at *industrial level*, in this section we will discuss the fuzzy control of a *switching converter*.

A DC/DC switching converter has the purpose of supplying a load with energy at constant voltage modulating the energy supplied at the source by storing it in a magnetic field and transferring it to an electric field that makes it available to the load. Figure 3.3 reports the basic topology of such converters.

They are characterized by high values of performance (up to and over 90%) and by the possibility of supplying an output voltage lower or higher than that of the input, depending on the topology adopted.

These converters have the task of keeping the output voltage constant under varying conditions of supply and/or load. Their regulation generally requires at least the feedback of the voltage error in order to modulate correctly the transfer of energy in every switching cycle. A switching cycle is characterized by two or three phases of operation according to the continuous or discontinuous operational mode:

ON phase: characterized mainly by the storing of energy in the magnetic field of the inductor.

OFF phase: characterized mainly by the transfer of energy stored in the inductor to the electric field of the condenser that feeds the load.

FW (free wheeling) phase: this is present only in *discontinuous* operation mode and is determined by the complete transfer of the stored energy to the inductor; this phase is characterized by the elimination of the inductor current that forces the electric field of the condenser to supply the load.

The *discontinuous* operational mode tends to be set up for low load currents and high feed voltage. Instead, in the *continuous* operational mode, the current in the inductor is never null; this means that, the load being equal, it presents a lesser ripple with respect to the discontinuous operational mode.

At all events, under conditions of equal output current, a converter in the continuous mode requires higher inductance values *L*. For the inductor, this means higher cost and greater bulk, greater loss in the magnetic core and a greater number of coils, and thus an increase in loss in the windings. The higher the switching frequency, the greater is the weight of this loss. In the continuous mode, as opposed to the discontinuous mode, the value of the output voltage is not connected with the output current value (load condition); that is undeniably an advantage of this mode of operation, in which, furthermore, the static gain in voltage of the converter proves to be higher and fewer loop gains are needed for its regulation.

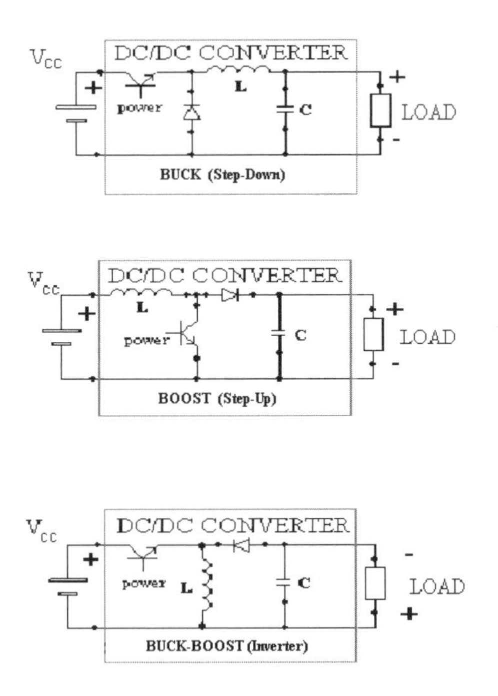

Figure 3.3. Main topologies of a DC/DC switching converter

Moreover, in the continuous mode, the fact that the inductor current is not null leads, in the transfer function, to the presence of complex and conjugate poles, and therefore to possible overshoot in response to the transistors, as well as to greater stability problems in the feedback loop. In particular, in the *boost* and *buck-boost* topologies in the continuous operational mode, the presence of partly positive zeros makes the feedback loop intrinsically unstable [1]. In a converter designed for operating in the discontinuous mode, it should be prevented, by suitably limiting the maximum peak of the inductor current, from passing into the continuous mode with the resultant risk of instability. Instead, for a converter designed for the continuous mode, passing over into the discontinuous mode may generate excess of voltage at the output (insufficient loop gain). When particular performances are required of a DC/DC converter in terms of noise, it is an advantage to adopt a control scheme which operates at fixed frequency, in which the regulation is achieved by modifying the ratio between the ON phase and the OFF phase, but not the overall duration of the commutation cycle. This type of regulation, Figure 3.4, is called *pulse width modulation* (PWM) at fixed frequencies and can be achieved by various methods which differ in the way in which the modulator relates the duty cycle to the voltage error and to the other control parameters, such as the inductor current, output current, *etc.* Generally, the voltage error determines the threshold with which a voltage ramp appears; once this threshold has been crossed, a latch needs to be set to pilot the power cut-out (end of the ON phase). An oscillator sets the commutation frequency, resetting the latch and commanding the switching on of the power. The ramp, synchronized with frequency, is generated in a *voltage-mode control* of the oscillator, whereas in a *current-mode control* it is obtained from the power current itself proportionally translated into voltage by a small R_{sense} resistance. Today, intense research and development activity is being carried out into DC/DC switching converters which aim at obtaining high performance both in terms of response to variations in load or feed voltage, and of imposing respect for increasingly tighter specifications regarding electromagnetic-type disturbance introduced by the converter. The problem of converter performance is particularly felt in those applications requiring high speed of response to variations in conditions of load and feed and a wide range of stability.

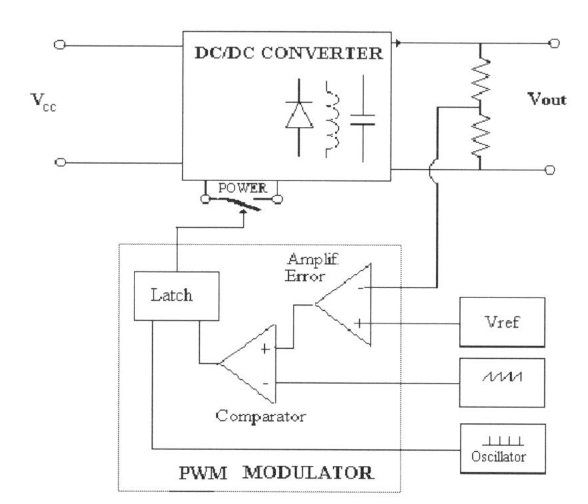

Figure 3.4. PWM control scheme

Various strategy have been proposed in order to overcome the difficulties connected with the use of linear controllers; new techniques have been proposed such as the *sliding-mode* [2], compensation ramps with variable slope [2], and more recently *fuzzy control* [4-7]. In particular, the advantages deriving from the use of fuzzy controllers have been pointed out, and especially those regarding *robustness* and *stability* over a wide range of operating conditions.

The interest in this non-linear control technique is justified in addition by the relative simplicity of controller synthesis. In fact, unlike other approaches, *a priori* knowledge can be transferred to the system in the form of *linguistic rules*. Further supports for the designer are the commercially available development programs which allow easy *tuning* of the *rule base*.

A *fuzzy control scheme* applied to the control of DC/DC converters is shown in Figures 3.5 and 3.6.

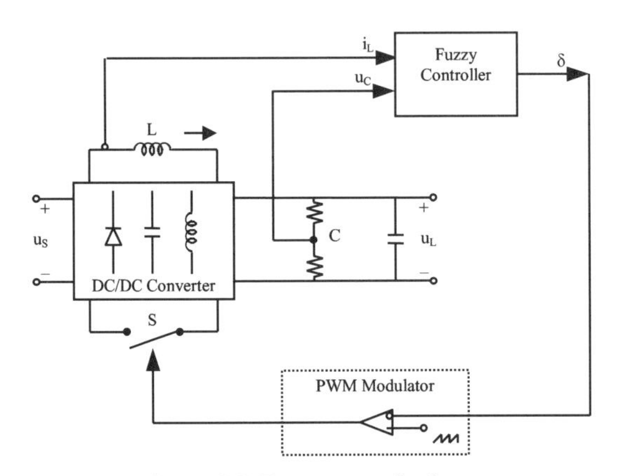

Figure 3.5. Fuzzy control scheme

In such a scheme, the inductor current and output voltage, together with their deviations with respect to the reference values (I_{LREF}, U_{OREF}), are employed as magnitudes in the input to a fuzzy controller which has the function of determining the duty cycle by means of a *proportional-integrative* action like that shown in Figure 3.6.

The rules can be derived on the basis of heuristic criteria such as those reported here below:

- when far from the s*et point*, ε_U is highly positive (*highly negative*). Strong corrective action: δ_P opposed to ε_U; in addition, to limit the presence of overshoot, $\Delta\delta_I$ is set close to zero;
- near the s*et point*: the error on the current is taken into account;
- if ε_U and ε_I are close to zero: δ_P and $\Delta\delta_I$ are set to zero (*regime condition*), and the duty cycle is determined by the integral term;
- if εu and ε_I are negative: with the aim of decreasing the system energy, δ_P and $\Delta\delta_I$ are set negative since both the output voltage and the inductor current are greater than the reference values;

- if ε_U and ε_I are positive: with the aim of increasing the system energy, δ_P and $\Delta\delta_I$ are set positive since both the output voltage and the inductor current are lower than the reference values;
- if ε_U is positive and ε_I negative (or *vice versa*): to avoid undershoot and overshoot, δ_P and $\Delta\delta_I$ are set to zero. Once, in the respective intervals of interest, the universe of discourse has been identified for the input (ε_U, ε_I, I_L) and output (δ_P, $\Delta\delta_I$) variables of the fuzzy blocks, it is then possible to determine the number and shape of the fuzzy sets over which the respective universes of discourse should be distributed.

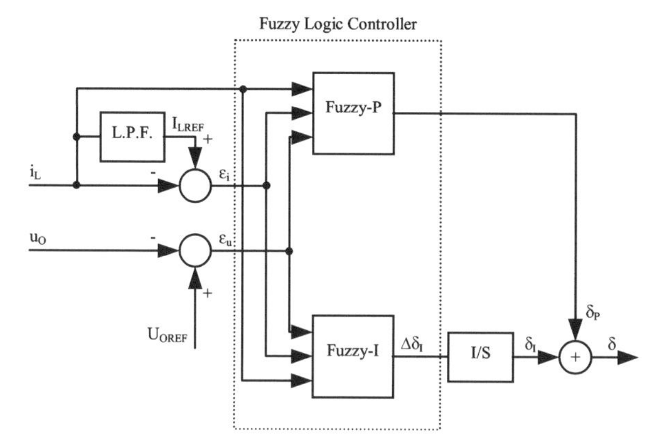

Figure 3.6. Fuzzy.Control

Five fuzzy sets are determined for the variables ε_U and ε_I, and two for the fuzzy variable I_L, while seven are determined for each of the output variables δ_P and $\Delta\delta_I$ (Mamdami algorithm); these memberships are summarized in normalized form in Figure 3.7.

The determination of 22 and 17 rules for the proportional and integrative fuzzy blocks, respectively, formulated on the basis of the previously mentioned criteria, allows us to make the desired control maps. In Figure 3.8, these control maps are represented and summarized by means of tables of implemented rules. The scheme proposed by the authors for a possible loop implementation and the results obtained in simulations are reported in Figures 3.9 and 3.10 respectively.

In the field of control, fuzzy logic affords the designer a structure which can implement a non-linear control characterized by a high degree of flexibility and generality; on the other hand, these characteristics demand a large number of parameters which have to be arranged by means of a tuning phase on the application or on models of it.

Undoubtedly, these characteristics are convenient for producing more suitable control schemes for overcoming the problems which may arise with traditional control, in those applications, where it is precisely the traditional techniques that are insufficient to guarantee the correct (or desired) behavior of the system.

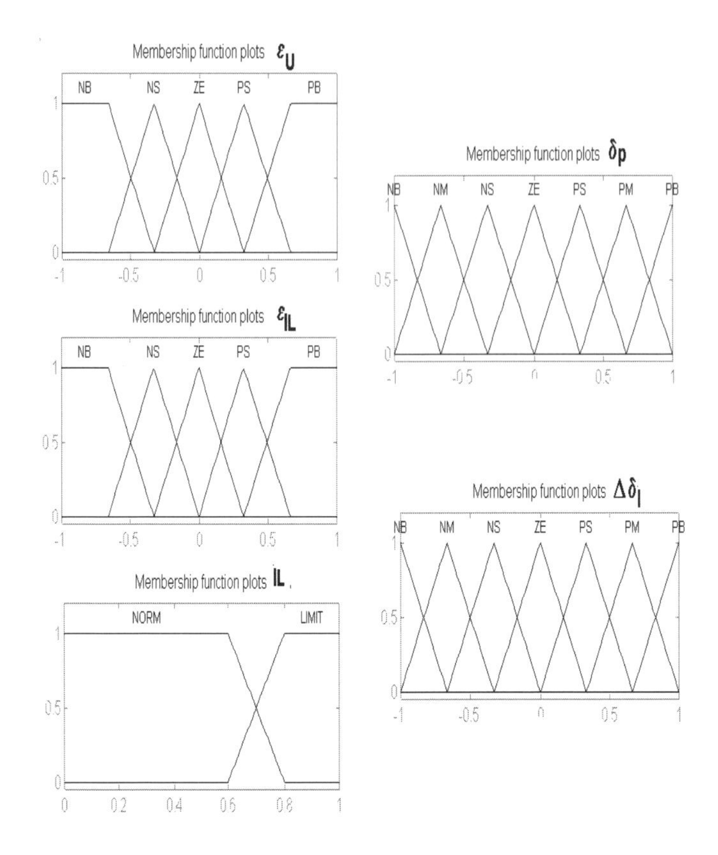

Figure 3.7. Fuzzy algorithm membership functions

On the other hand, the number of parameters and the possible interactions triggered by the definition of the various memberships tend to make every application of fuzzy logic a case study in which empirical knowledge of the process in question is fused with that of this kind of control technique. From this standpoint, every application achieved by means of fuzzy logic aims at demonstrating, or at least verifying, the best performances attainable with respect to traditional control or the indispensability of this technique in specific applications.

To the present authors, it seems quite normal that, by increasing the number of parameters employed in the control, one can obtain a system behavior characterized by better performance. But the increase in inherent complications due to fuzzy logic being implemented means that it must be demonstrated that these complications be justified in terms of results obtained.

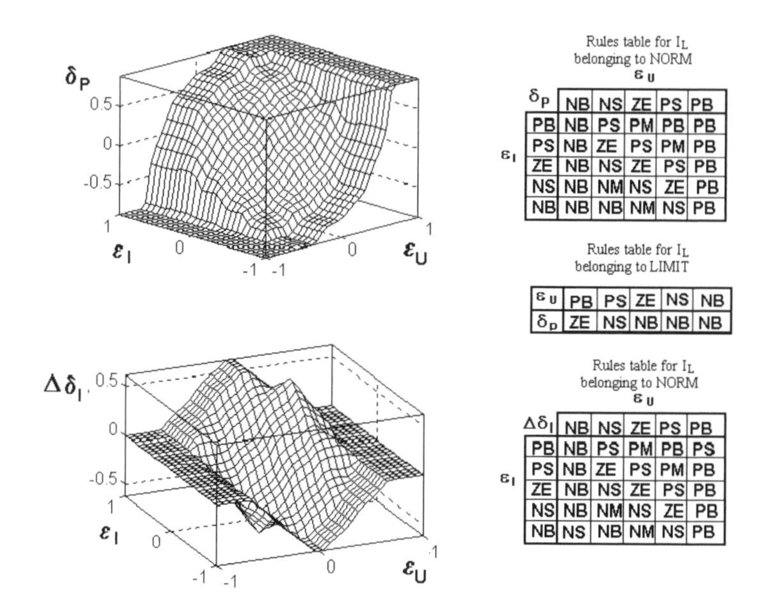

Rules table for I_L belonging to NORM

δ_P \ ε_U	NB	NS	ZE	PS	PB
PB	NB	PS	PM	PB	PB
PS	NB	ZE	PS	PM	PB
ZE	NB	NS	ZE	PS	PB
NS	NB	NM	NS	ZE	PB
NB	NB	NB	NM	NS	PB

(rows: ε_I)

Rules table for I_L belonging to LIMIT

ε_U	PB	PS	ZE	NS	NB
δ_P	ZE	NS	NB	NB	NB

Rules table for I_L belonging to NORM

$\Delta\delta_I$ \ ε_U	NB	NS	ZE	PS	PB
PB	NB	PS	PM	PB	PS
PS	NB	ZE	PS	PM	PB
ZE	NB	NS	ZE	PS	PB
NS	NB	NM	NS	ZE	PB
NB	NS	NB	NM	NS	PB

(rows: ε_I)

Figure 3.8. Control maps and rule base

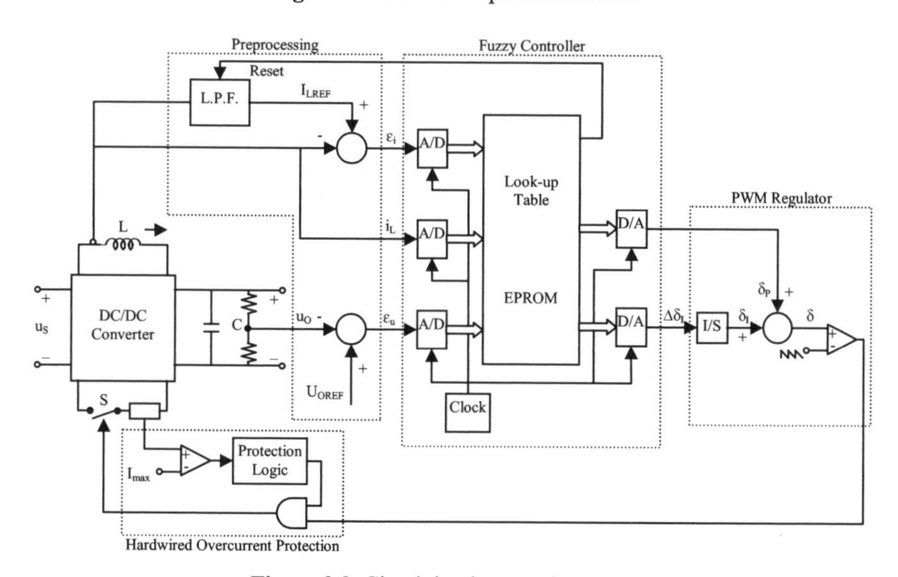

Figure 3.9. Circuit implementation scheme

In the field of DC/DC converters, traditional control in the range of the field of application affords good regulation at very low cost.

Applications requiring an extensive range of stability, correct behavior both in the continuous and discontinuous mode, and "intelligent" management of the anomalous conditions cannot do other than fall back on control techniques that afford an adequate and different action according to the various situations, and fuzzy logic is one of these.

However, these applications do not present that aspect of generality that will increase the interest in fuzzy logic and they contribute very little to creating the *accessibility* that is needed for forming fuzzy logic *customers*. In addition, the difficulties and complications in solving them, due to the almost total lack of products for implementing such techniques, put fuzzy logic even further back into the field of particular applications and *case studies*. These difficulties can be overcome by using appropriate *processors* and *micro-controllers*, as is described in the following example. This is a solution particularly indicated for those applications in which several co-acting fuzzy algorithms are used.

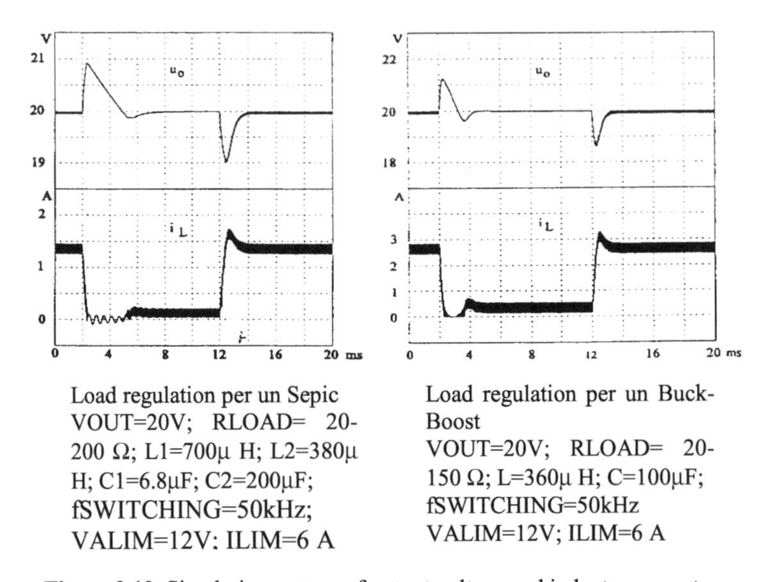

Load regulation per un Sepic VOUT=20V; RLOAD= 20-200 Ω; L1=700μ H; L2=380μ H; C1=6.8μF; C2=200μF; fSWITCHING=50kHz; VALIM=12V: ILIM=6 A

Load regulation per un Buck-Boost VOUT=20V; RLOAD= 20-150 Ω; L=360μ H; C=100μF; fSWITCHING=50kHz VALIM=12V; ILIM=6 A

Figure 3.10. Simulation pattern of output voltage and inductor current

The *ST52E301 micro-controller* is suitable for making an evaluation chart for the development of fuzzy controllers. The block diagram is reported in Figure 3.11.

This device began life as a member of the *STMicroelectronics* fuzzy processor and co-processor families *W.A.R.P.1* and *W.A.R.P.2* (Weighted Associative Rule Processor), and is characterized by the possibility it offers of implementing Boolean and fuzzy algorithms.

The component in question is among other things equipped with 2 kbytes of EPROM and a series of peripherals that give it considerable I/O flexibility.

The *fuzzy core* includes a fuzzifier, a unit of inference, and a defuzzifier and is able to manipulate about 300 rules (characterized by 4 inputs and 1 output) that can be shared by several fuzzy subroutines and activated by conditions defined by the user.

The device interface is achieved by means of the following units that are present inside it:

- four A/D converters;
- an I/O parallel port;
- an I/O serial port;

- a programmable timer;
- a Triac/PWM Driver, which allows power devices to be piloted directly.

The technical characteristics of this component and the discretion connected with its use present certain features that cannot really be considered those most suitable for controlling switching converters, which, instead, need rapid controllers capable of not introducing an appreciable delay between their input and output.

It has been decided to make this technique accessible and therefore to use this component, trying, with the smallest possible number of components, to conciliate the characteristics of the micro-controller with the requirements of the application.

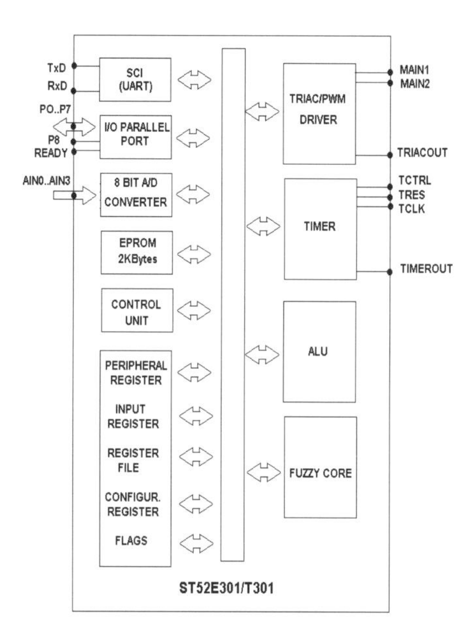

Figure 3.11. Block diagram of ST52E301

The purpose is in this way to obtain a better adaptation to the load conditions of the converter, making it independent of the operational mode (continuous/ discontinuous).

Thus we describe the possibility of obtaining through fuzzy logic an error amplifier capable of associating the regime condition with those of the error and load.

It can be observed that the derivative of the output voltage error can be adopted as an expression of the load conditions; that is justified by the fact that during the ON phase of the converter in the boost and buck-boost configurations, only the output condenser sustains the load current at the expense of the stored load.

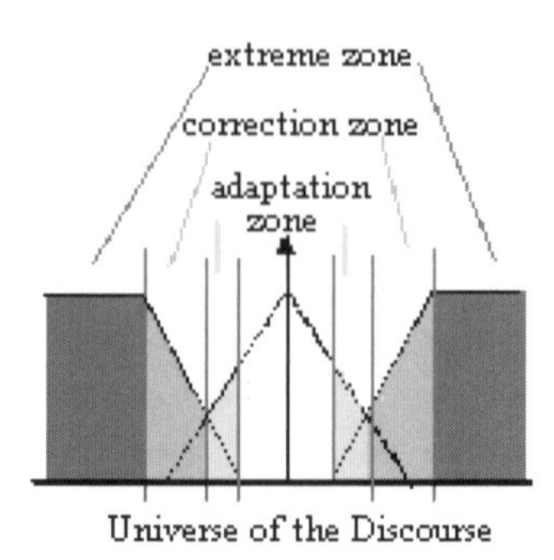

Figure 3.12. Control zones

Neglecting the drop on the resistance, the derivative of the output voltage (or of its error) proves to be proportional to the load current. Therefore the error derivative can be employed for determining the load conditions to which the fuzzy algorithm output may be modulated if the output voltage error is close to zero. In conditions of error other than zero, that is to say in conditions *external* to that of the regime, zones in need of heavy control action can be identified.

This choice means that no distinction should be made, merely on the basis of voltage error, between the *extreme* operational conditions such as a *short-circuit* (which will be fed by determining the maximum duty cycle) and *no-load operation* (in which the converter will tend to switch off determining the minimum duty cycle).

It can be hypothesized that as the *regime conditions* are approached, the extreme operational conditions must be attenuated and combined in order to determine the exact value of the duty cycle associated with the specific operational condition.

This law of attenuation and combination can be expressed by means of the shape of membership, while the regime value (in conditions of slight error) is determined by the whole fuzzy algorithm.

A further hypothesis regarding the fuzzy algorithm, but one limited to the memberships of the error variable, is the need for every action to present an error zone in which it is pre-eminent, that is to say a negative error expressing a lower output to the one desired imposes a vaster duty cycle, whereas a positive error requires a diminution.

This hypothesis implies a law of variation of the control action connected in almost linear fashion to the error; in this way, in the interests of simplicity, it is possible to avoid to consider a more sophisticated structure which would be able to take into account also non-linear behaviors of the converter such those present in switching from continuous to discontinuous operational mode and *vice versa*.

With this sort of hypothesis, such aspects which are solved by the fuzzy algorithm by means of the values assumed by the second fuzzy variable, the derivative of the error, are not taken into account in the fuzzy error variable.

With regard to the error variable, the determination of two *extreme* zones defines a whole zone of regulation in which the zero error condition is located.

A certain degree of indetermination of this condition can be easily obtained through a zero error membership with which an *approximate* and accepted error zone can be identified.

In this zone, the error is close to zero and it is precisely the degree of membership to this condition that determines an *adaptation* to the regime condition imposed by the second variable. Another four zones can be identified in addition to this one. A first zone in which, apart from the zero error membership, also one of the two *extreme* membership conditions is active.

Adjacent to this, a further major *correction zone* of the control action can be identified in which the degree of extreme membership activation prevails over that of the zero error.

A total of seven zones with which the control action is connected to the output error can be identified as depicted in Figure 3.12.

In the case in which it is desired to make the error variable contribute to the identification of anomalous operational conditions, for example that of a short circuit, only very few modifications need be made to the previously defined fuzzy variable, adding to it a further membership connected to the short-circuit condition.

In this way, in order not to consider the short-circuit condition as merely one of crossing a threshold, a zone is identified in which the short-circuit condition tends to be confused with that of overloading. It is important to note that in this way zones (error belts) are defined, characterized by *behavioral* aspects (such as overloading and short circuit), which are increasingly more pre-eminent inside the identified zones.

In the case of the DC/DC converters, the error variable alone is not enough to reveal a short-circuit condition, since, on the basis of the error, this can be confused with the start-up condition.

A knowledge of the sign of the error derivative allows the two situations (*i.e.*, an increasing error due to cto cto, and a decreasing one due to the start up) to be distinguished.

In the application, the maximum circulating current should be limited by switching off the converter immediately whenever the current rises above the maximum (threshold) value allowed by the components.

For this reason, in the example proposed, the control over the maximum current is implemented by means of traditional techniques.

Considerations analogous to the previous ones are at the base of any determination of the memberships connected with the second fuzzy variable.

For this purpose, zones are identified in which it is found to be *pre-eminent* to attribute a certain value to the fuzzy block output.

A tuning phase of these memberships allows the fuzzy algorithm to be defined.

In the example proposed, a *density* of the memberships is noted with the increasing of load conditions.

The decision to use the *ST52E301 micro-controller* and the conversion times of its analog-to-digital converters do not allow the error variation to be used as a significant expression of the load conditions; it is thus essential to identify another magnitude which is representative of the transfer of energy to the load.

The *scheme* adopted for this is the one proposed in Figure 3.13, where, to determine the regime working point, the output voltage error and power current are taken into account. The latter was chosen for its being a representative magnitude of the transfer of energy to the load.

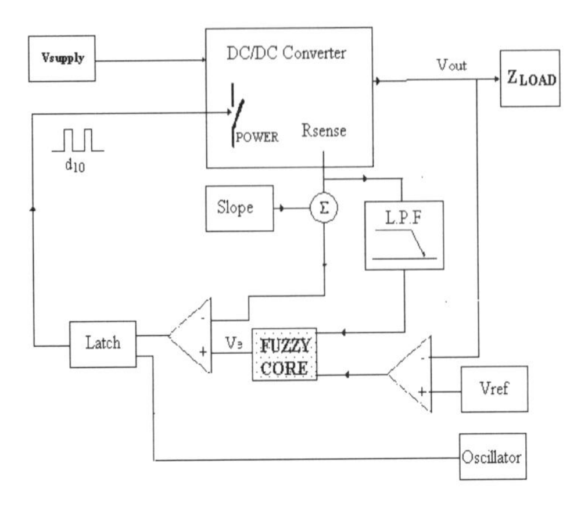

Figure 3.13. The operative scheme adopted

In fact, although other magnitudes may represent load conditions, the error derivative assessed during the ON phase, they do not agree very well with the requirements of the conversion frequency. In fact, in every switching cycle of the converter, one reading only is made. In the scheme of the application, it is therefore essential to introduce *low-pitch filters* (*cf.* the L.P.F. and INTERFACE blocks in Figure 3.14 below) which have at least the same time constant as the system, in order to determine the averaged values in the switching cycle and the significant control readings.

In fact, Figure 3.14 represents the *circuit diagram* of the scheme proposed for the case of a Boost configuration converter.

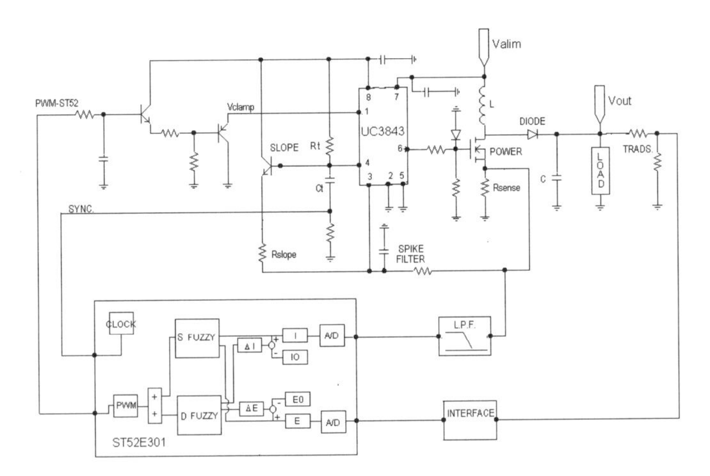

Figure 3.14. Loop scheme for the example proposed

The *ST52E301 component* is used as a fuzzy controller in the voltage loop, while the *PWM driver*, acronym UC3843, is the current loop as seen in the scheme depicted in Figure 3.14.

In fact, the use of a special micro-controller for implementing soft computing techniques together with a traditional controller for switching converters proves to be a solution characterized by a high degree of flexibility and economy.

The power current is used for recognizing the set point since, as has already been pointed out, it plays a significant role in the transfer of energy from the magnetic field of the inductor and the electric field of the condenser. During the transients, this magnitude is unfortunately not representative of the load conditions; therefore, also the derivative of this current must be used in addition to that of the output voltage error.

Basically, the fuzzy core, Figure 3.16, consists in two *blocks*, respectively known as *S-fuzzy* and *D-fuzzy*. The former is preferred for maintaining the *regime conditions* in function of the load and feed conditions, while the latter block has a *derivative-type* action and is thus dedicated to improving the stability.

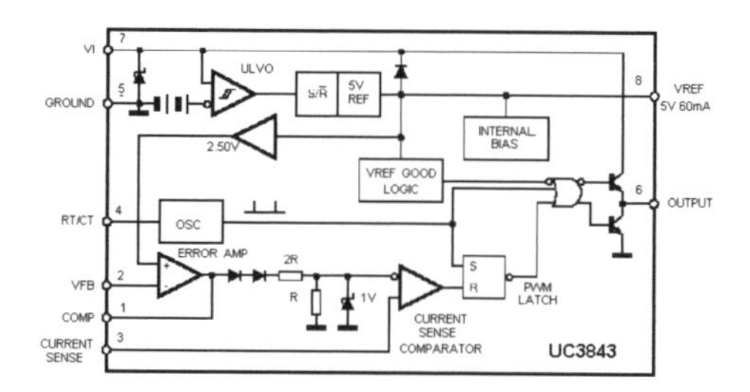

Figure 3.15. UC3843 block diagram

The intensity of the *D-fuzzy* block is determined on the basis of variations in the output voltage ΔE and in the average current in the power device in the IPM commutation cycle.

With respect to the transient response, it should be noted that the control action is complicated by the natural delay introduced by the presence of the A/D conversion organs and by the digital processor that can be assessed together in two clock cycles.

In the scheme considered, the output voltage is compared with an internal reference, determining the error with respect to the set point.

To the point in which it is representative of the energy transfer, the power current contributes to determining the desired information regarding the operational conditions.

The information about the feed voltage value is implicitly present in the current-mode type of control scheme, in which the signal on the PWM modulator ramp is proportional to the inductor current that flows into the power during the ON phase. That is to say it is proportional to the inductor current during the load phase, and the current slope is seen to be connected with the input voltage.

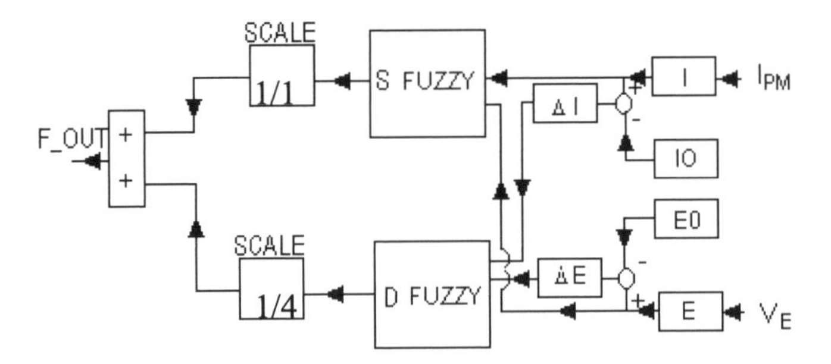

Figure 3.16. Fuzzy core block diagram

The fuzzy core output is applied to one of the two inputs of the comparator making up the *PWM modulator* and governing the switching on and off of the power transistor (duty cycle).

The purpose is to act in a non-linear fashion on the converter transfer function, as shown in the basic scheme reported in Figure 3.14.

The *S-fuzzy* block has the task of determining the F_OUT regime value and is based on a set of heuristic-type linguistic rules according to the following criteria:

- for a *small* absolute value error, the fuzzy block output (*'F_OUT'*) is determined on the basis of the load condition (*'Ipm'*);
- a *high positive* value of error forces F_OUT up to its maximum value;
- a *high negative* value of error brings F_OUT down to its minimum value.

The *D-fuzzy* block, whose action is felt only during any transients, is also heuristically derived and has the purpose of introducing into F_OUT a contribution which is opposed to the variations appearing in the transient, with the aim of setting the system in regime conditions.

In the scheme reported in Figure 3.16, the two linguistic input variables in the *S-fuzzy* block are:

- the output vltage error V_E: *i.e.*, the difference between the output voltage and the reference one;
- the average intensity of the power current in an I_{PM} cycle, in so far that it is an expression of the transfer of energy by the load converter.

By means of the analog-to-digital conversion operating inside the *ST52E301*, the universe of discourse for each of the variables V_E and I_{PM} is brought to the range [0, 255] by use of an *8-bit A/D converter*.

For the V_E variable, three fuzzy sets were fixed, labeled *Eneg, Ezero*, and *Epos*, respectively. For the current variable, instead, five fuzzy sets were identified and labeled, respectively: *Very Small* (VS), *Small* (S), *Medium* (M), *Large* (L), *Very Large* (VL).

By means of a *trial and error* procedure based on the use of the type of converter already described, the membership functions were tuned.

The structure of the rules is that proposed by Sugeno with the consequent values as constant rules, *i.e.* expressed by real numbers and not by fuzzy variables.

The position of the membership functions in the respective universe of discourse and the control maps are reported in Figures 3.17 and 3.18 for *S-fuzzy* and *D-fuzzy* blocks, respectively.

When the error signals are equal, the *S-fuzzy* block supplies output values which depend in non-linear fashion on the regime operating conditions of the regulator: in this way, high loop gain values can be obtained for the regulator, but only in zones of intrinsic static stability (the presence of a positive zero), and without making use of low-frequency poles. These would penalize the transient behavior of the regulator and at the same time be in contrast with the start of oscillation.

For high error values, the shape of the *S-fuzzy* block membership functions can minimize the dependence of the loop gain on load and feed conditions, thereby determining a greater speed of response to the load and feed transients, but however maintaining the regulator in the stability zone. The rule base for the S-fuzzy block consists in seven rules identified on a purely empirical basis.

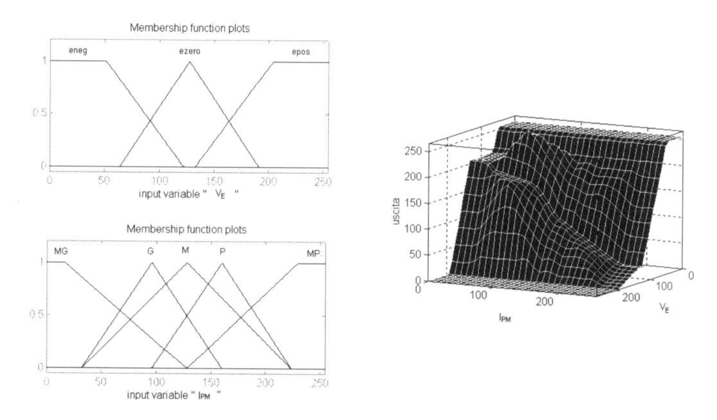

Figure 3.17. Membership functions and control map for an S-fuzzy control block

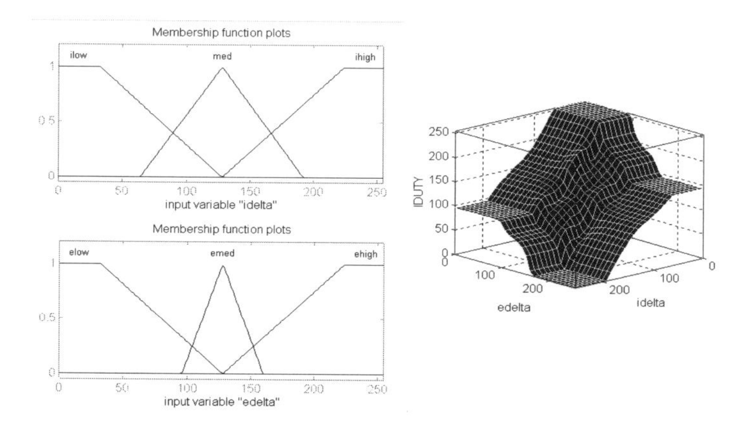

Figure 3.18. Membership functions and control map for a D-fuzzy control block

The *D-fuzzy* block was developed with a view to generating an antagonistic action with respect to the oscillations that arise during the transients, both on the power current and on the output voltage. There are two variables present in the part of the rule base antecedents of this block: the variation of the output voltage error, hereafter called *edelta*, and variations in the mean power current, hereafter called *idelta*. For each of these variables, three fuzzy sets have thus been identified: *elow*, *emed*, and *ehigh* for the voltage, and *ilow*, *imed*, and *ihigh* for the current. These variables are calculated inside the micro-controller and their respective universes of discourse lie between 0 and 255. The rule base for the *D-fuzzy* block, whose structure is still that of the type proposed by Sugeno, thus consists in nine rules, as shown schematically in Table 3.1 below. Each *entry* in the Table identifies a rule, and the value reported for it is the intensity of the derivative action performed by the *D-fuzzy* block. The range [0, 255] should be considered as divided into two equal parts to allow negative and positive values of the derivative action to be represented.

Table 3.1. The rule base identified for the D-fuzzy block

	Elow	emed	Ehigh
ilow	255	192	144
med	160	127	96
ipos	96	64	0

With the purpose of illustrating the overall operation of the converter in question, Figure 3.19 reports the patterns of the inductor current and output voltage with varying feed (from 9 to 12 V) and load (200, 15, 45 Ω) for a DC/DC boost converter ($V_{OUT} = 15$, $L = 170\ \mu H$, $C = 470\ \mu F$, $F_{SWITCH} \approx 14.285$ kHz, $I_{MAX} = 3$ A) obtained in SIMULINK® environment simulation. The system can be seen to operate well, both in the continuous ($IL \neq 0$ in the cycle) and discontinuous mode within a considerable range of load condition variations.

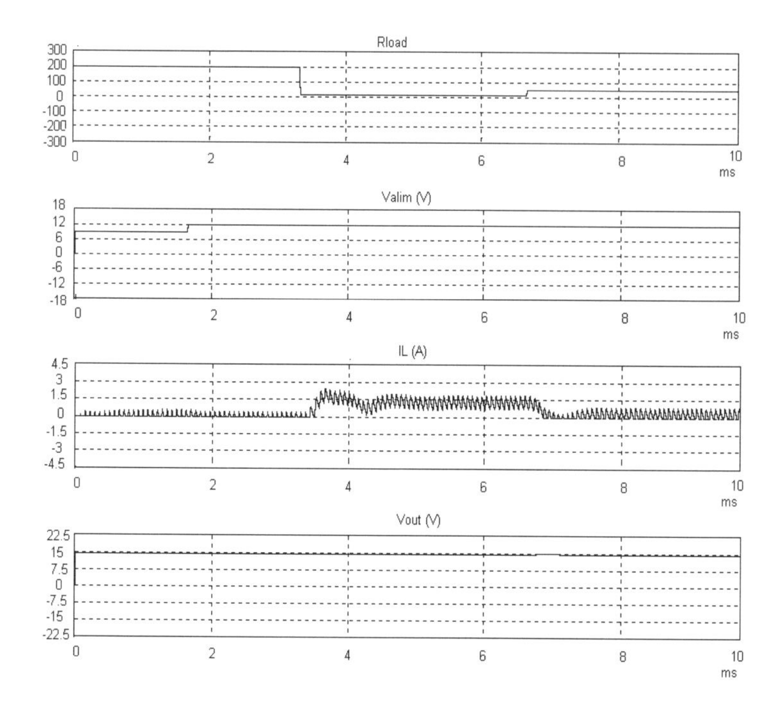

Figure 3.19. Output current (I_L) and voltage *Vout*) with variation in the voltage (*Valim*) and load (*Rout*)

The example offered shows very clearly how the design of a fuzzy controller cannot be accomplished in the automatic mode, using exclusively the techniques proposed, but in any case requires to be integrated with the knowledge of an expert, and the use of the classical strategies and optimization strategies that from time to time are considered indicated. However, what has been said is a characteristic of all soft computing techniques, that derive their efficacy precisely from their possibility of being integrated and possibly modified to adapt to the characteristics of the system that is to be controlled. Another example of fuzzy control, determined by using genetic algorithms as an optimization technique, will be reported in Chapter 11.

3.5 References

1. Pozzolo V. Gli Alimentatori Stabilizzati a Commutazione. Dipartimento di Elettronica–Politecnico Torino, Italy
2. Mattavelli P, Rossetto L, Spiazzi G. Small-Signal Analysis of DC/DC Converters with Slide Mode Control. IEEE Transactions On Power Electronics. 1997; 12: 1
3. Dong Tan F. Current-Loop Gain with Nonlinear Compensating Ramp. Proc. of 27th Annual Power Electronics Spec. Conference, Baveno, June 1996; 23-27

4. Criscione M, Lionetto A, Nunnari G, Occhipinti L. Embedded Fuzzy Control on Monolithic DC/DC Converters. Proc. of IEEE ISCAS. 1998
5. Consoli A, Oriti G, Scarcella G, Testa A. Fuzzy Control of a Fast High-Quality Rectifier. in Proc. of 27th Annual Power Electronics Spec. Conference, Baveno, June 23-27, 1996
6. So WC, Tse CK, Lee YS. Development of Fuzzy Logic Controller for DC/DC Converters. IEEE Transaction on Power Electronics. 1996; 11: 1
7. Mattavelli P, Rossetto L, Spiazzi G, Tenti P. General-purpose Fuzzy Controller for DC/DC Converters. IEEE Transactions on Power Electronics. 1997; 12: 1
8. Fortuna L, Graziani S, Baglio S, Nunnari G. Improvements in fuzzy controller design. Proc. Of the IEEE Int. Conf. on System Engineering, Ohio, 1991; 237-240
9. Fortuna L, Muscato G, Nunnari G, Occhipinti L. Neural Modelling and Fuzzy Control: Regulation of the Temperature of an oven. Journal of Systems Engineering, Springer Verlag, London UK, 1995; 5: 61-75
10. Fortuna L. Gallo A, Vinci C, Xibilia M.G. Modelling of a Fuzzy Controller for a Single Link Manipulator. Journal of System Analysis Modelling Simulation, Gordon and Breach Science Publ., 1993; 13: 247-254

4. Artificial Neural Networks

4.1 Introduction

As has been seen previously, fuzzy logic arose from the attempt to *emulate* the imprecise and incomplete way of treating information that is typical of the *human brain*. Another attempt in this direction is represented by *neural networks*, which were born as neural structure models of the brain but are currently used as *calculation paradigms* for *classifying* and *approximating* non-linear functions.

One of the characteristics of neural networks is that of emulating the structure of the human brain, even though somewhat simplistically, by exploiting its main characteristics, *i.e.*, the ability to learn from experience [1-3], [7-8].

This arises from the consideration that the human brain, although slower and less precise than more sophisticated systems of digital calculation, is able to resolve more efficaciously certain problems such as recognizing images, reconstructing them on the basis of noise data, and in general all problems of classification [5], [11-12], [25-29], [33], [41], [43].

Although the functioning of neural biological systems is still insufficiently understood, it is known to be based on the *parallel action* of a large number of simple operations such as the transmission, or not, of a signal, according to the *stimuli* they receive from other neurons.

A basic characteristic of the operational mechanism of brain tissue is the so-called *learning through examples*, which is to say that every new experience contributes to increasing the knowledge acquired, thereby improving performance.

Thus, neural networks emulate the structure and functioning of biological neurons, even if in fact no neural network is able to reach, except to a minimal extent, the elaborating capacity of the human brain, and therefore the analogy is a purely formal one.

Unlike the usual systems of digital calculation, which need a program in order to perform the operations on the input data required of them, neural networks need a *training phase* to acquire, by means of the *examples* presented to them, the necessary *experience* for supplying correct output when faced with new input.

In general terms, an artificial neural network can thus be defined as a tightly connected *computational system* able to *store* and *utilize knowledge* acquired through experimenting. The knowledge acquired is stored with the help of values of certain parameters, called *weights*, which connect the computational units, known as *nodes* or *neurons*, whose values are fixed during the training phase. Every neuron is an entity which has access to several inputs and one output only. It receives the inputs from *neighboring* neurons, elaborates them and transmits the output to other neurons, appropriately weighting them by means of the connective values. The operations performed by the neurons are generally simple ones of *addition*, *non-linear mapping*, or *thresholds*. The neuron model currently most used is reported in Figure 4.1, where x_i represents the *i*-th input (the *i*-th component of the vector X), w_i is the relative weight at the *i*-th input (the *i*-th component of the vector W), and $f(W^tX)$ is a function, generally a non-linear one, known as the *activation function* [1]. The value of the activation function, calculated at the weighted addition of the inputs, thus represents the *neuron output*. The neural networks currently in use are distinguished according to the nature of the operations performed by the neurons, *i.e.*, according to the nature of the activation function and to the way in which the neurons are connected [1-3], [8].

The choice of structure is heavily dependent on the application considered.

Independently of the different types of neurons and the structure of the net, two sets of neurons can generally be identified: the so-called *input neurons*, which have the task of receiving the inputs from the net and transmitting them to the other successive neuron layers, and the set of neurons forming the *outputs* of the net. The data supplied to the input are therefore transmitted to the other units, elaborated, and transmitted to the outside by the output neurons.

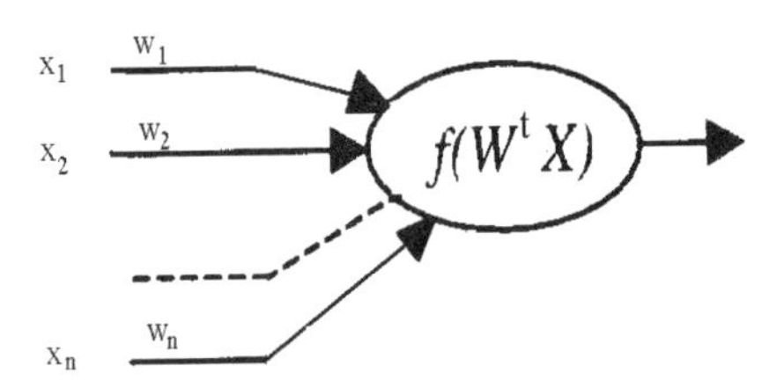

Figure 4.1. Neuron model

where:

$$X = [x_1, x_2, \ldots x_n];$$

$$W = [w_1, w_2, \ldots w_n].$$

Whereas the number of input and output neurons is known once the application has been fixed, the *hidden structure* is chosen from one time to another and markedly determines the *performance* that can be attained with the net. Since an excessive number of nodes can compromise the correct functioning of the net, it is generally preferred to start off with networks having a reduced number of neurons, and gradually to increase the number until the desired performance is reached. In the majority of cases, there are three different *operational phases* of the networks. The first is called the *learning or training phase*. During this phase a set of data, representing the knowledge base of the considered problem, are provided to the network.

These data, using different learning algorithm, choosen according to the network structure and type, are used to update the weight values.

The learning algorithms are subdivided into *supervised* and *non-supervised* algorithms. In particular, they are called supervised when during the learning phase an input-output set of data is fed into the network. For each of the inputs, the net calculates the corresponding output and compares it with the *correct* one which has been attributed to it. The algorithm determining the value to attribute to the weights generally performs a *minimization* of the standard deviation between the network output and that desired for all the data supplied during the training, which are presented to the net in iterative form until an acceptable error is reached. If the error obtained is not satisfactory and does not decrease with an increasing number of iterations, the network topology will have to be modified by increasing the number of nodes.

Instead, the learning algorithms are said to be *non-supervised* when the correct outputs are not fed into the net. Generally these algorithms are used for dividing the inputs into distinct *classes*, which are identified by a different the salient features of the elements of each class. The second phase is called the *test phase* or *checking* phase and, by using a different data set, serves to verify that at the end of the learning the network is functioning correctly. In this phase, the weights of the net are kept fixed at the previously determined values and one merely has to verify that the error obtained with the new set of data is still sufficiently low. If that is not the case, the structure of the network may not be correct, or the data fed in during the learning may be insufficiently representative of the problem, or the number of free parameters to be determined (weights) is excessive with respect to the number of data supplied during the learning. In the last of these cases, the number of input-output data supplied must generally be increased or else, if it is insufficient, the number of parameters be reduced. Reduction of the number of parameters is achieved by choosing a topology with a smaller number of nodes, even if that generally implies a worsening of performance during the learning; therefore, a satisfactory *compromise* must be found. Once these two phases are completed, the net can be used for calculating the output in the face of *unknown inputs* (those not presented during the learning phase); that is to say, it behaves as a calculation algorithm in order to solve the problem for which it was trained. Classical applications of neural networks are classification, reconstruction of data disturbed by noise, approximation of non-linear functions, prediction of time series, optimization and control [4-6], [11-12], [25-28], [31], [33], [41], [43].

To tackle the problems of *classification*, input values defining an element belonging to a given class are fed into the net, and the net is asked to learn the characteristics distinguishing this element and to identify the correct membership class.

Instead, if the problem is *approximation* of a non-linear function, the input values used are those belonging to the domain of the function, and, after having learnt during training the mechanism which generated the data, the net is required to calculate correctly the corresponding values of the function. Other problems frequently solved by neural networks are those concerned with the *reconstruction* of incomplete data or of data disturbed by noise. For these applications, the net is supplied during training with a set of disturbed data together with the exact datum to which they correspond. The network is therefore required to *associate* with every input the corresponding disturbance-free value. In the following sections, the most common neural network topologies and their learning algorithms will be described, and the main applications and their limitations, if any, be outlined. In particular, the *multilayer perceptron*, the *Kohonen nets*, and the *radial basis function network* will be analyzed.

4.2 The Multilayer Perceptron

The first neural structure to be proposed in the literature was the well-known *perceptron* [7-8]. Rosenblatt's perceptron will not be analyzed in detail on account of the intrinsic limitations in its structure, which make it unsuitable for solving problems of classification where the classes are not *linearly separable*. That is to say a set of inputs activating a given output value, identifying its membership to a class, can be separated by means of a hyperplane from the set of inputs belonging to another class.

An example of a non-linearly separable classification problem is that of the XOR logical function (exclusive OR), defined by the following table:

Table 4.1. XOR logical function

input 1	input 2	output
0	0	0
1	0	1
0	1	1
1	1	0

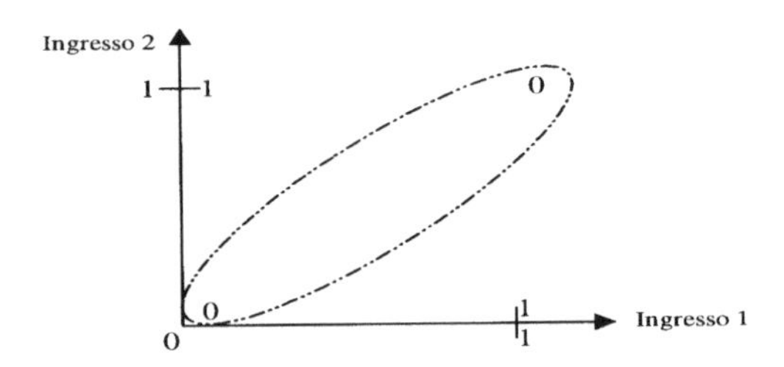

Figure 4.2. XOR input-output function map

Let us suppose that the elements giving the output 0 belong to the class 0, and that those giving the output 1 belong to the class 1, and considering the function to be represented on the input 1 – input 2 plane, as in Figure 4.2, it is seen that the two classes cannot be separated by a line.

This limitation was overcome by the neural network known as the *multilayer perceptron* (MLP) [1], [13], [36].

As shown in Figure 4.3, the multilayer perceptron is made up of a layer of input neurons that merely have the task of transferring the signal to the next layer, giving it the appropriate weights from a certain number of internal layers, also known as *hidden layers*, and from a layer of output neurons. The neurons of each layer are all connected by the *weights* with those of the previous and successive layers. The applicational potential of the MLP derives precisely from the presence of the internal layers. For reasons that will be explained later in greater detail, the most frequent configuration is that with one internal layer only [9-10], [42].

The MLP structure is thus characterized by the *number of hidden layers* and the *number of neurons* in each layer. Even when limiting the number of cases to that of a network with only one hidden layer, there is no rule for determining the correct number of hidden neurons with respect to the problem considered. Generally, as already mentioned in the introduction, it is preferred to start from a network with few internal neurons, and then progressively increase the number until good performances are achieved. One can also proceed in the opposite direction, starting from a network with a high number of internal neurons and progressively eliminating those that make less contribution to determining the correct I/O ratio [16-17], [19], [20], [40].

With regard to Figure 4.3, the following notations are used:

- $N1$: number of layers;
- s=1,...,N1: index of layers;
- $N^{(s)}$: number of neurons in the layer s (in particular $N^{(1)}$ indicates the input while $N^{(N1)}$ indicates the number of output);
- $O_i^{(s)}$: i-th input of the layer s (for s=1 it represents the i-th input of the network);
- $W_{ij}^{(s)}$: weight of the interconnection between the i-th neuron of the layer s and the j-th neuron of the layer s-1;
- $\theta_i^{(s)}$: the bias factor of the i-th unit of the layer s, which corresponds to the weight of an interconnection with a neuron activated at the unitary value;
- $\sigma i(\bullet)$: the activation function (generally, a continuous and monotone increasing derivable function, bounded both above and below).

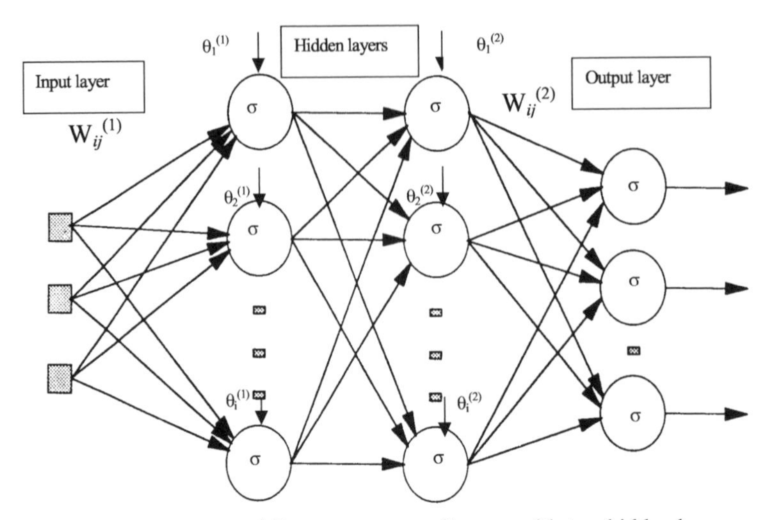

Figure 4.3. Multilayer perceptron diagram with two hidden layers

The formulas which allow us to calculate the network outputs, once the inputs $O_i^{(1)}$ are known, are the following:

- calculation of the activation for each neuron

$$x_i^{(s)} = \sum_{j=1}^{N_{s-1}} W_{ij}^{(s-1)} O_j^{(s-1)} + \theta_i^{(s)} \qquad \text{for } s=2,\ldots, N_e \; e \; i=1,\ldots,N^{(s)} \tag{4.1}$$

- calculation of the neuron output with a logistic sigmoidal activation function

$$O_i^{(s)} = f(x_i^{(s)}) = \frac{1}{1+e^{-x_i^{(s)}}} \tag{4.2}$$

The calculations that allow us, given the input values to the network, to calculate the outputs $O_i^{(NI)}$ for $i = 1,\ldots,N^{(NI)}$ are generally called *forward propagation* phase.

Since in this way, due to the considered activation functions, the network outputs assume values ranging from 0 to 1, the data fed into the net must be suitably *normalized* in all applications. Generally, in order to obtain better performances, the outputs are normalized in such way as to bring them into the interval [0.1 0.9]. Alternatively, without in any way compromising the performances that can be theoretically obtained by the network, a *linear activation function* can be considered for the network output layer; this is generally equal to the identity function, and thus normalization of data can be avoided.

The rapid success enjoyed by this artificial neural network model and the corresponding learning logarithm is for a vast part linked to the theoretical results obtained on the basis of the Stone-Weierstrass theorem which offers a rigorous demonstration of the *universal interpolation* property of a MLP with at least one layer of hidden neurons. In fact, it has been proved that a MLP-type artificial neural network with at least three layers and a *sigmoidal activation* function for the neurons of the hidden layer is able, to *any degree of accuracy* that may be required, to interpolate a continuous non-linear function, or one with a finite number of discontinuities [9-10]. However, this theorem does not offer any result with regard to the number of neurons needed to obtain the required accuracy.

A classical example able to highlight the potentiality of the MLP is the previously mentioned problem of the XOR.

As is shown in Figure 4.4, a three-layer MLP having a 2-2-1 topology can separate the input space into two non-linearly separable decisional regions. It can be noted that in such a case, there are two input neurons and one output neuron because the law to be approximated is $X^2 \rightarrow X$ with $X = \{0,1\}$, while the number of internal neurons has been obtained experimentally starting from a 2-1-1 topology and increasing the number of internal neurons until good performances were obtained.

The functional potentiality of a MLP can mainly be set in two different applicational frames. The first frame is connected with capacity of implementing non-linear transformations in order to solve *functional approximation* problems, while a second group of applications consists in the possibility of automatically *partitioning* the space of the patterns and thus solving problems of *classification* having decisional regions of any form whatsoever.

As will be seen in the following chapter, also the problems of *identification*, *prediction* of temporal series, and non-linear system *control* can be brought back to problems of non-linear function approximation; to train a MLP in such way as to be able to simulate dynamic systems or temporal series, it will be sufficient to add to the input layer further units having the task of memorizing one or more samples of the network outputs at the previous instants [11-12].

With regard to the *learning* of a MLP, it takes place by means of an *iterative algorithm* that updates the values of the network interconnections such that a total *square error functional* is optimized on a set of input/output data defined as

$$E = \tfrac{1}{2}\sum_{p=1}^{npat} E_p = \tfrac{1}{2}\sum_{p=1}^{npat}\sum_{i=1}^{nout}(ydes_{i,p} - ycalc_{i,p})^2 \tag{4.3}$$

where, modifying the notation in the interest of simplicity, *npat* is the number of data supplied during learning, *nout* is the number of outputs, *i.e.*, of neurons in the last layer, *ydes* is the desired output value, and *ycalc* is the corresponding value calculated by the forward propagation phase of the network. With every iteration, one of the training set samples is presented to the neural network. After the stimulus corresponding to the input vector has been forward propagated, the error committed by each neuron output, given by $ydes_{i,p} - ycalc_{i,p}$, is calculated.

At this point, the gradient algorithm for *back-propagation* of the output layer error is applied until the first hidden layer, simultaneously updating the value of the weights, according to the *deepest-descent* gradient formula [13]:

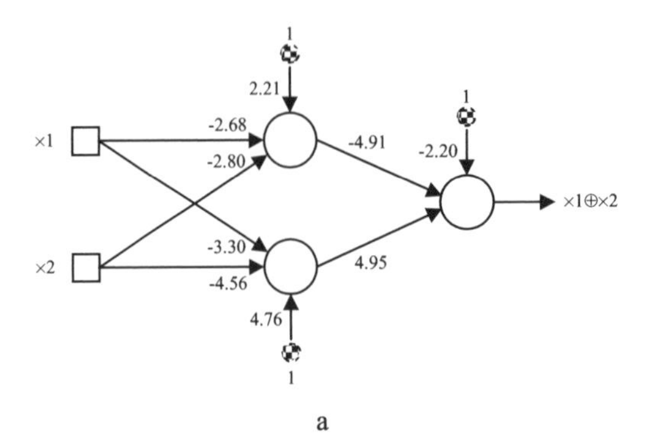

a

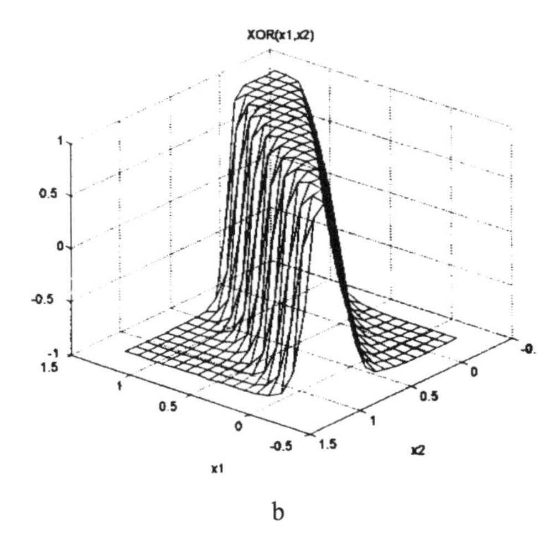

b

Figure 4.4. a. XOR multilayer perceptron; b. formation of the decisional regions in the XOR

$$\Delta W_{ij}^{(s)}(t) = -\varepsilon \frac{\partial E_p}{\partial W_{ij}^{(s)}} \tag{4.4}$$

where ε is a parameter chosen by the user and known as the *speed of learning*. Updating the weights proportionally opposite the error gradient with respect to the same weights, we in fact obtain the minimization of the error with respect to the weights. By calculating the error functional gradient in the case of a MLP, we obtain the following set of equations that make up the so-called *delta rule*, better known as the *back-propagation error algorithm*. Adding a further second order term, the so-called *momentum* (μ), which takes the learning history into account, has the effect of increasing the speed of convergence of the algorithm.

$$\Delta W_{ij}^{(s)}(t) = -\varepsilon \delta_i^{(s)} \cdot o_j^{(s-1)}(t) + \mu \Delta W_{ij}^{(s)}(t-1) \tag{4.5}$$

$$W_{ij}^{(s)}(t+1) = W_{ij}^{(s)}(t) + \Delta W_{ij}^{(s)}(t) \tag{4.6}$$

where:

t = the index of iteration;
$\delta_i^{(s)} = f'(x_i^{(s)})\ (ydesi - ycalci)$;
for the output layers, if the activation function is sigmoidal;
$\delta_i^{(s)} = f'(x_i^{(s)})\ \Sigma_k W_{ki}^{(s+1)} \cdot \delta_k^{(s+1)}$;
for the hidden layers.

For the previously chosen activation function, the calculation of the derivative $f'(x_i^{(s)})$ is very simple; in fact, we have:

$$f'(x_i^{(s)}) = f(x_i^{(s)})\ (1\text{-}f(x_i^{(s)})).$$

To train the network, we can proceed in two different ways. The first technique, known under the name of training *by pattern*, foresees the following steps:

1. initialization of the weights;
2. presentation of a sample in the input;
3. forward propagation phase for the output calculation;
4. comparison between the calculated output and the desired output;
5. back-propagation phase of the error and calculation of δ;
6. updating of the weights.

The steps are repeated, starting from the second, until all the training samples have been presented.

The iteration is repeated on all the training samples and is given the name of *epoch*.

The training of the network proceeds by effecting a number of epochs such as to obtain a sufficiently low error or until the error no longer decreases, which indicates the incapacity of the network to solve the considered problem. The other type of training, called training *by epoch*, consists instead in updating the values of the weights only at the end of every epoch, and accumulating during the epoch the increase of the weights calculated on every sample.

With regard to the choice of values of ε, it should be noted that too small a value slows down the convergence while too great a value can lead to error oscillations during the convergence training. The addition of momentum, corresponding to the insertion of a low pass filter in the minimum search directions contributes to reducing the oscillations. The values of ε and μ are generally modified during the learning [14-15], [18]. The validity of the back-propagation algorithm has been widely proven by the numerous applications both in the industrial field and in the purely scientific one.

The main problem of using such an algorithm is, apart from the often excessively low speed of convergence, the possibility of obtaining configurations of the weights corresponding to a local minimum. The literature reports numerous variations of this algorithm, aimed at improving the performance [8]. Besides, global optimization algorithms, such as the genetic ones, are also frequently used for solving the problem of *local minima*.

As has already been pointed out, the good performance of a MLP depends not only on a correct choice of the data to be used for the learning, but also on the use of an *adequate topology*.

The remarks made previously can therefore be summarized as follows: too small a network will not, in general, be able to yield a good model of the problem; the more the problem is a complex one, the more the number of neurons in the hidden *layer*(*s*) will have to grow in order to achieve the correct interpolation function. It should also be noted that although one single hidden layer is theoretically enough, the use of networks with several hidden layers sometimes allows the total number of units to be reduced.

On the other hand, too large a network can implement numerous solutions consistent with the training data of the problem, the majority of which, however, achieve a bad approximation of the real problem, being limited to a mere association between the input data and the output data. These networks are thus unable to respond correctly to the various data presented to them during training, in which case one can say that the network does not have a good capacity of *generalization*. For this reason, and possibly by using *a priori* knowledge of the process, *trial and error* types of procedure are employed; these consist in starting from a topology with a small number of neurons in the hidden layer, and gradually increasing the complexity of the net until the optimal topology is reached.

A method for determining whether the topology obtained is the optimal one is *cross-validation*, which consists in making available, together with the training set, another set of input/output data forming the *test set*. During each cycle of the learning process a set of training data is given to the network. Once the output error is computed it is back propagated for updating the interconnection weights; at the same time, the network outputs, and then the errors, are calculated on the test set data. By comparing the *mean square error* made on the training set (LMSE) with the corresponding one on the test set (TMSE), the emergence of *overfitting* can be evidenced, which occurs when the TMSE is markedly higher than the LMSE.

Another method for obtaining optimal MLP topology is to start from a sufficiently large network and successively eliminate the units or interconnections that contribute in negligible fashion to the functioning of the network. The principle on which this technique is based can be found in the literature [19-20].

It should be noted in any case that, since from the theoretical standpoint the network can solve the problem with the desired accuracy, if no good results are obtained with any topology, that depends on the bad choice of samples supplied during the training. In fact, the necessary condition at the base of the neural approach to problems of classifying and approximating functions is that the data available be sufficient in number and sufficiently representative of the problem in question.

4.3 Non-supervised Neural Networks

Non-supervised neural networks are generally used for solving problems of *classification* [21]. For these networks, the learning algorithm tends to group the input samples on the basis of common characteristics that are not explicated by a *correct output*. Rather, what is carried out is a regrouping of the data without having any *a priori* knowledge of whether an input sample belongs to a particular class. A sample not introduced during the learning will be associated with the most similar of those added to the network, on the basis of a *measure of similarity* that could be, for example, distance or the inner product.

One of the possible neural structures utilized for this type of learning is that made up of only two layers [1], [3], [8]. The input layer has a number of neurons equal to the size of the input vector, *i.e.*, to the size of the element that must be classified. The input units transmit the datum to the output units by means of weights. The output units are equal in number to the number of *classes*, which in general is not precisely known *a priori*, and carry out a simple *weighted sum* of the inputs without using activation functions. The *direct propagation phase* is thus very simple. With reference to Figure 4.5, we can define the following notations:

- N: number of input neurons;
- M: number of output neurons;
- u_j: j-th output;
- x_i: i-th input;
- w_{ij}: the weight connecting the i-th input with the j-th output.

It is generally assumed that the inputs are normalized in such way that every input vector is of unitary length. An analogous normalization is performed during training for the vector of the relative weight at each output.

Therefore, for calculating the j-th output, we have:

$$u_j = \sum_{i=1}^{N} w_{ij} x_i \tag{4.7}$$

The most common learning rule is known as *winner takes all*. The algorithm can be schematically expressed as follows:

1. an input sample $P \in R^N$ is introduced into the net;
2. the output for each neuron is calculated;
3. the output with the greatest activation value is selected, *i.e.*, the output u_m which has won the competition;
4. only the vector of the weights corresponding to the unit u_m is updated so as to reduce the distance between the vector of the weights and the sample introduced into the input; thus, the following rule is used:

$$\Delta w_{im} = \varepsilon(p - w_{im}) \quad i = 1,...N \tag{4.8}$$

At the conclusion of the learning, all the weights that connect the input units to the j-th output neuron will represent the center of the j-th class in the N-dimensional space of the inputs.

ε is the speed of learning which is chosen heuristically, and is progressively diminished during the learning. The updating of the weights is carried out for each input sample and the iterations are repeated until the variation in weights obtained at every iteration becomes negligible. Immediately after updating, the weights are again normalized.

In this case, too, the learning can be affected by accumulating the variation of the weights and updating them at the end of the epoch.

The choice of a correct size for the output layer is given by the fact that all the neurons prove to be activated for at least one, or a certain number of, input samples.

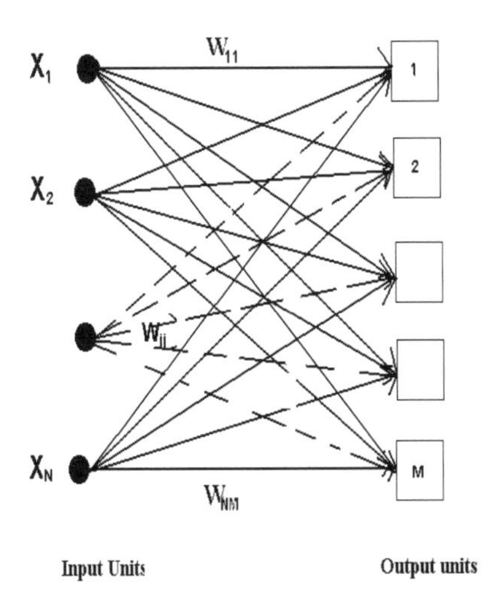

Figure 4.5. Competitive-type training network

Once the training has been completed, the network can be used for *classifying* the samples not used during the learning. The membership class of each sample is identified by the output neuron that presents greatest activation.

Although the *winner takes all* algorithm offers considerable potential, it does not achieve a correct mapping when the data characteristics are not immediately apparent or when the boundaries between the classes are not clearly distinct.

An important modification of this type of non-supervised network that, while conserving the same structure, allows better performances to be achieved, is to be found in the *Kohonen self-organizing networks* [22], [30]. In these structures, the arrangement of the neurons inside the layer assumes importance. The learning process leads to the formation of topological maps in which the most important relations of *similarity* encountered among the input samples are converted into *spatial relationships* among the neurons that respond to these inputs.

The learning algorithm of the Kohonen self-organizing maps differs from the *winner takes all* in the way in which the weights are updated. The basic difference is that not only the weights of the *winning* neuron are updated, but also those of the *spatially close* neurons, on the basis of a law that establishes the zone of activity for each neuron. An example of a Kohonen network with 2 inputs and 49 outputs is reported in Figure 4.6.

The same figure indicates the possible zones of activity with radius 3, 2, and 1. The neighborhood radius is generally decreased during the training. The learning algorithm is similar to the previous one, the only difference being that the weights relative to all the neurons belonging to the interior are updated.

$$\Omega_c(i) = \exp\left(\frac{-\|u_c - u_i\|^2}{2\sigma^2}\right) \tag{4.9}$$

where u_c is the output of the neuron having the greatest activation, the index i varies over all the units of the network, u_j is the output of the j-th unit, and σ is a parameter characterizing the surroundings. The updating is therefore achieved with the law:

$$\Delta w_j = \varepsilon \;\Omega_c(i)(p - w_j) \tag{4.10}$$

where p is the input sample for all the units.

The value of σ generally decreases with an increase in the iterations, as does that of ε.

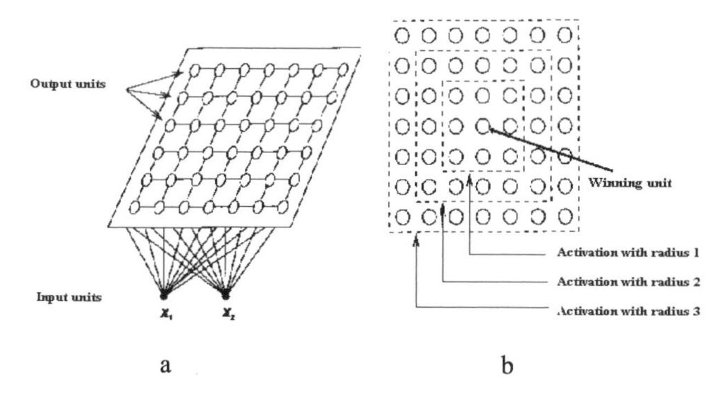

Figure 4.6. a. Kohonen self-organizing network with 2 inputs and 49 outputs; b. Activation zones with radius 3, 2, 1 around the winning unit

4.4 Radial Basis Function Neural Networks

A considerable disadvantage of MLP is the high non-linearity in the parameters. To determine the parameters characterizing the network it is necessary to adopt non-linear optimization techniques whose use leads to the risk to remain trapped inside local minimum configurations.

The *radial basis function* (RBF) neural networks [23-24] represent an alternative to the perceptron type in approximating non-linear I/O maps starting from examples and in classification.

The architecture of a RBF network is made up of an input layer whose neurons have only the function of transferring the signal to the successive stage by means of unitary value weights, from one layer of the hidden neurons and from an output layer.

The hidden neurons form zones of *localized receptivity* with respect to the input stimuli by means of an *activation function*, e.g., a *Gaussian* one. This means that the hidden neurons produce a response other than zero only when the input vector of the unit is found inside a limited input space region, identified by the activation function of the neuron.

The output layer performs a linear combination of the output of the hidden neurons, whose coefficients are the weights connecting the hidden layer with the output layer. Once the activation functions parameters have been fixed, the RBF network is thus linear in its parameters.

This kind of structure arose from emulation of some receptive structures present in the cerebral and visual cortex.

The name *radial basis function* originated from the fact that the Gaussian functions have radial symmetry. The difference with MLP is also due to the fact that the sigmoidal functions form a base that assumes non-zero values in an infinitely vast region of the input space while, as has already been said, the Gaussian functions cover only small zones of the inputs.

Whereas some functions are better approximated by linear combinations of sigmoids, for others, especially classification problems, better results are obtained with the RBF.

As will be seen in what follows, the main advantage of RBF networks is their *reduced computational load*, both in the training phase and when processing data, although from the theoretical viewpoint they have the same properties as MLP ones.

That occurs generally at the expense of lesser accuracy of approximation and a lesser capacity to generalize with respect to MLP.

With reference to Figure 4.7, the following notations are adopted:

- n: number of inputs;
- n_r: number of hidden neurons;
- n_o: number of outputs;
- n_p: number of samples per training;

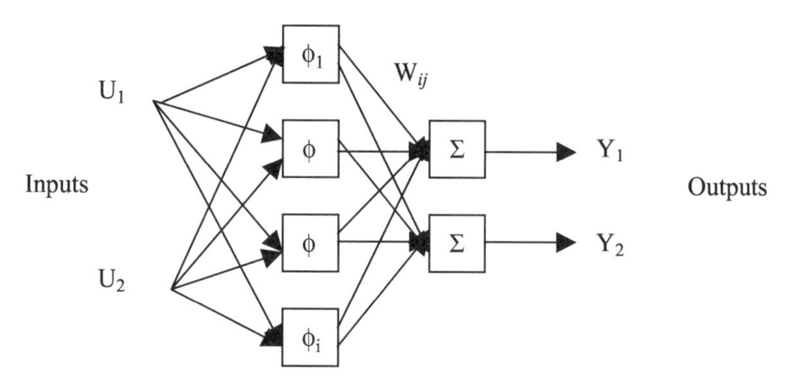

Figure 4.7. RBF neural network scheme

- $d \in R^{no}$: k-th desired output (k=1,...n_p);
- $x_{(k)}$: = $[x_{(k)1} ... x_{(k)n}]$: input vector;
- $\phi(\bullet)$: activation function of the hidden neurons;
- $w_{ij} \in R$: the weight connecting the i-th output unit with the i-th hiddenneuron ($w_j = [w_{ij},...w_{nj}]$);
- $C_i \in R^n$: (i=1,...,n_r) center of the activation function;
- $\sigma_i \in R$: variance of the i-th activation function;
- y_i: j-th output of the network (j=1,...,n_o);
- $u_i \in R$: i-th output of hidden neurons (i=1,...n_r).

The forward propagation phase of the inputs is described by the following relations (the index k has been omitted for the sake of simplicity):

$$u_i = \phi\,(\|x\text{-}C_i\|)$$

$$y_j(x) = \sum_{i=1}^{n_2} w_{ij} u_i \tag{4.11}$$

$$= \sum_{i=1}^{n_r} w_{ij}\,\Phi(\|x - C_i\|) = \sum_{i=1}^{n_r} w_{ij} \exp\left(-\|x - C_i\|^2 / 2\sigma_i^2\right)$$

$$\text{per} \quad j = 1,\dots,n_o$$

As can be seen from Relation 4.11, the output $\Phi(\|x - C_i\|)$ of the i-th internal neuron is only activated if the input is close to the center C_i, the size of which is determined by the *variance* σ_i.

The centers of the activation functions are usually chosen in the set of input points.

Apart from the Gaussian activation function, others can be chosen without modifying the network's performance [23].

A large number of *training algorithms* exist for RBF networks, most of which separate the training into two distinct phases, *i.e.*, that of *choosing the centers* and variances, and that of *calculating the weights*.

The *training of the hidden layer*, that is to say the choice of the layers, is generally performed in a *non-supervised* fashion, whereas that of the *output layers* is *supervised*.

The algorithm originally developed for training the RBF networks requires that there be a number of RBF *centers* equal in number to the points of the function to be approximated; however, generally this number is very large, which makes the network structure excessively complex.

Sometimes, the centers are chosen randomly, but this cannot be considered a valid criterion for setting up RBF networks, since all too often it leads to a bad numerical conditioning due to some centers being too close.

A very widely used algorithm for choosing the centers is the so-called *k-means algorithm*, which is a very simple *clustering* algorithm capable of yielding good results.

The *k-means algorithm* proceeds according to the following steps:

1. initialization of the C_j $j=1,\dots,n_r$ *centers* (generally these are chosen equal to the first n_r training samples) and clustering of all the data around the nearest center;
2. for all samples of input x_i $i=1,\dots,n_p$, a center is selected to which they are nearest; to do that, for all the centers the value of $j = \| x_i - c_j \|$ is calculated, and the vector x_i is assigned to the center C_j for which the distance of j is minimal;
3. calculation of the data averages:

   for all the centers, we calculate $C_j = \frac{1}{M_j} \sum_{x_i \in I_j} x_i$ where I_j is the set of all the data previously clustered around the center C_j, and M_j is the number of data in I_j.

Steps 2 and 3 are repeated until there is no change in the clustering of the data from one iteration to another.

Once the centers have been calculated, the variances, which represent a measure of the data dispersion around the centers, are generally calculated as

$$\sigma_j^2 = .\frac{1}{M_j} \sum_{x_\in I_j} (x_i - C_j)^T (x_i - C_j)$$

After the training of the hidden layer has been completed, by means of this algorithm, it is possible to proceed to calculate the weights of the output layer; this is generally done with the *least squares* (LMS) algorithm [35], [36]. The data set needed for the training is formed by the couples (*w*, *d*) where the values of *w* are calculated for a certain set of data using the values of the centers and the previously determined variances, and the values of *d* are the corresponding output.

The algorithm which minimizes the standard deviation between the network outputs and the values desired for all the training samples is the following: for each of the n_o outputs (j=1,...,n_o) we build the vector $d_{Mj} \in R^{np}$ containing all the j-th components of the desired outputs for all the samples $d=[d_{(1)j}...d_{(k)j}...d_{(no)j})]^T$ and the matrix

$$U_M = \begin{bmatrix} u_{(1)1} \dots u_{(1),n_r} \\ u_{(k),1} \dots u_{(k).nr} \\ u_{(np),1} \dots u_{(np),nr} \end{bmatrix}$$

containing all the outputs of the hidden units for each input sample;
the vector w_j of the parameters is calculated as:

$$w_j = (U_M^T U_M)^{-1} U_M^T d_{Mj}$$

and thus it is seen that the calculation of the weights for the last layer does not require an iterative algorithm.

Another method for selecting the centers is that where each center corresponds to a given regressor in a *linear regression model*; therefore, the choice of the RBF centers can be looked on as one being made by means of the *orthogonal least squares (OLS) method* [23]. With this method, the bad conditioning problems and those regarding a very large number of centers can be avoided.

Relation 4.11 can be considered a particular case of the linear regression model:

$$d(k) = \sum_{i=1}^{M} p_i (k)\, \theta_i + \varepsilon (k) \qquad (4.12)$$

where $d(k)$ is the desired output, θ_i the parameters, $p_i(k)$ the regressors which are the functions of $x(k)$, and $\varepsilon(k)$ is the error and is presumed to be not correlated with the regressors.

Each center C_i with a certain non linearity $\Phi(\bullet)$ corresponds, in (4.11), to a regressor $p_i(k)$, and thus the problem of center selection can be seen as a problem of choosing significant regressor subsets.

The learning algorithm used in [22] can be derived from the OLS method.

To apply the least squares method to the RBF network, (4.12) can best be written in a matrix form, *i.e.*:

$$d = P\Theta + E \qquad (4.13)$$

where:

$d = [d(1)...d(N)]^T$;

$P = [p_1..p_M]$, with $p_i[p_1(1)...p_i(N)]^T$ for $1 \le i \le M$;

$\Theta = [\theta_1...\theta_M]^T$;

$E = [\varepsilon(1)...\varepsilon(N)]^T$.

The regressor vectors p_i form a base and therefore by applying the linear least squares (LS) method, an estimate $\hat{\Theta}$ for the parameters vector of (4.13) can be determined which will satisfy the condition that $P\hat{\Theta}$ is the projection of the vector d on the space covered by the base, that is to say it is a part of the desired output energy [23].

To understand how each regressor contributes to the output energy, the regression matrix P is decomposed into the product of 2 matrices:

$$P = WA \tag{4.14}$$

where:

A is a upper triangular matrix $M \times M$:

$$A = \begin{bmatrix} 1 & \alpha_{12} & \cdots & \cdots & \alpha_{1M} \\ 0 & 1 & \alpha_{23} & \cdots & \alpha_{2M} \\ 0 & 0 & \ddots & \ddots & \vdots \\ \vdots & \vdots & \ddots & 1 & \alpha_{M-1,M} \\ 0 & 0 & \cdots & 0 & 1 \end{bmatrix} \tag{4.15}$$

W is the matrix $N \times M$ with *orthogonal* columns w_i, for which we have:

$$W^T W = H = \mathrm{diag}\left(\sum_{i=2}^{N} w_i(t) w_i(t)) \right) \; 1 \le i \le M \tag{4.16}$$

Relation 4.13 can be rewritten in the following way:

$$d = Wg + E \tag{4.17}$$

The solution of (4.17) is obtained by applying the OLS method, for which we have:

$$\hat{g} = (W^T W)^{-1} W^T d \tag{4.18}$$

The vectors $\hat{g}$ and $\hat{\Theta}$ are joined by the relation:

$$A\hat{\Theta} = \hat{g} \tag{4.19}$$

Different ways exist to determine the regression matrix P, among which there is the *Gram-Schmidt* method [23]. To select the RBF centers subset from the set of data, usually a very large one, the OLS method is used [22]. The number M of significant regressors calculated with the OLS method is usually much smaller than N, and thus there is a considerable reduction in the computational complexity of the regression model parameters, and hence of the RBF network. Since w_i and w_j are orthogonal for $i \neq j$, the desired output energy $d(t)$ is:

$$d^T d = \sum_{i=1}^{M} g_i^2 w_i^T w_i + E^T E \tag{4.20}$$

Considering the vector d given by the difference between the desired output and its mean value, the variance of $d(t)$ can be expressed as follows:

$$N^{-1} d^T d = N^{-1} \sum_{i=1}^{M} g_i^2 w_i^T w_i + N^{-1} E^T E \tag{4.21}$$

where the first term of the second member represents the variance of $d(t)$ owing to the regressors and the second term expresses the unexplained (owing to error) variance of $d(t)$.

A simple yet efficacious way of choosing the subset of significant regressors is that of determining the relation of error reduction due to the i-th regressor which is defined in the following way:

$$err_i = \frac{g_i^2 w_i^T w_i}{d^T d} \qquad 1 \le i \le M \tag{4.22}$$

The procedure adopted for selecting the regressors, described in [23], exploits the Gram-Schmidt method and is the following:

- 1st step

$$\mathrm{w}_1^{(i)} = \mathrm{p}_i$$

$$g_i^{(i)} = \frac{(w_1^{(i)})^T d}{((w_1^{(i)})^T w_1^{(i)})} \qquad 1 \le i \le M$$

$$err_1^{(i)} = \frac{(g_1^{(i)})^2 ((w_1^{(i)})^T w_1^{(i)}}{d^T d}$$

The maximum of error is therefore sought whose index is indicated by i_1, and the first regressor is selected:

$$w_1 = w_1^{i1} = p_{i1}$$

- k-th step

$$\alpha_{jk}^{(i)} = \frac{w_j^T p_i}{w_j^T w_j}$$

$$w_k^{(i)} = p_i - \sum \alpha_{jk}^{(i)} w_j \qquad k \ge 2 \quad 1 \le i \le \mathrm{M}$$

$$g_k^{(i)} = \frac{(w_k^{(i)})^T d}{((w_k^{(i)})^T w_k^{(i)})} \qquad i \neq i_1, \ldots, i \neq i_{k-1} \qquad 1 \leq j < k$$

$$err_k^{(i)} = \frac{(g_k^{(i)})^2((w_k^{(i)})^T w_k^{(i)})}{d^T d}$$

The maximum of error is therefore sought, indicated by i_k, and the k-th regressor is selected:

$$w_k = w_k^{(i_k)} = p_{ik} - \sum_{j=1}^{k-1} \alpha_{jk}^{(i_k)} w_j \qquad 1 \leq j \leq k$$

The procedure finishes when:

$$1 - \sum_{j=1}^{M_g} err_j \langle \rho \qquad \text{with} \quad 0 < \rho < 1 \tag{4.23}$$

where ρ indicates the tolerance chosen.

In this way, one obtains a model containing M significant regressors.

4.5 Applications

In this section, two classical examples of artificial neural network applications are described. The first concerns the *interpolation*, by means of a multilayer perceptron, of a non-linear function having two variables. The second regards the use of RBF networks for *classifying* defects on board railroad trains.

4.5.1 Example: Interpolation of a Non-linear Map

Let us consider the following non-linear function in the space x, y, z:

$$z = xe^{(x-y)} = y*\sin(2x\pi) \tag{4.24}$$

The goal is to interpolate this function using a neural network.

The neural network considered is of the perceptron type with one hidden layer only, sufficient for the theorem of Cybenko [37] for solving the problem in question, and a sigmoidal activation function [38-39]. The configuration of the chosen network is described in the following table:

Table 4.2. Configuration of a network

Input neurons (x, y)	Hidden layer neurons	Output neurons	Learing pattern	Test pattern
2	12	1	800	100

The patterns used in the training phase, carried out by means of the *back propagation* algorithm, were obtained by giving the input variables x, y random values with Gaussian distribution in the interval [-1, 1]; that produces output values in the interval [-2, 8]. Needless to say, the patterns used in the test phase are distinct form those used during the learning phase.

The choice of the number of neurons in the hidden layer was made in the following way.

A relatively small number of neurons was considered and the mean square error was calculated for the test patterns (TMSE) after a sufficient number of cycles for stabilizing the error. Then the number of neurons in a unit was increased and the training was repeated. This procedure was carried out iteratively until the TMSE reached a value considered acceptable.

The number of neurons in the hidden layer at the start was set equal to 5, whereas the final number obtained by applying the procedure above was 12.

After the learning phase, the mean square error on the learning patterns (LMSE) and that on the test patterns (TMSE) assumed the following values, respectively:

$$LMSE = 0.00144$$

$$TMSE = 0.00145$$

Once the learning phase was completed, the interpolation procedure of the considered map was completed and the validation phase began. This consisted in generating the output values z corresponding to input couples (x, y) on a uniformly distributed grid of points in the interval [-1, 1] with step 0.1.

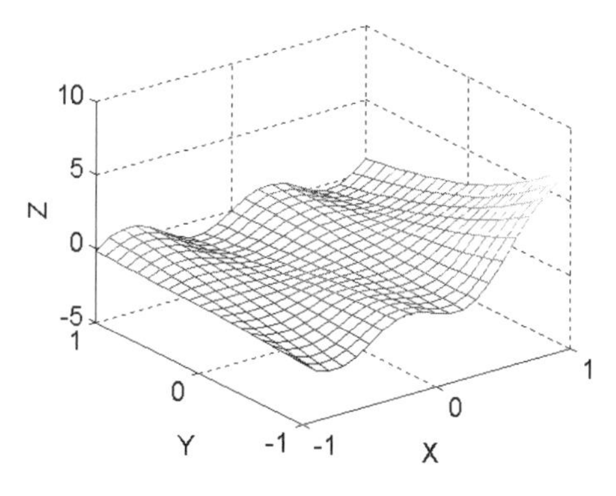

Figure 4.8. Map of true values

Figure 4.8 shows the map obtained in the region considered using the "true" values calculated by means of (4.24). Figures 4.9 and 4.10 show, respectively, the interpolated map obtained using the neural network output values and the map of difference between the true values and the interpolated ones.

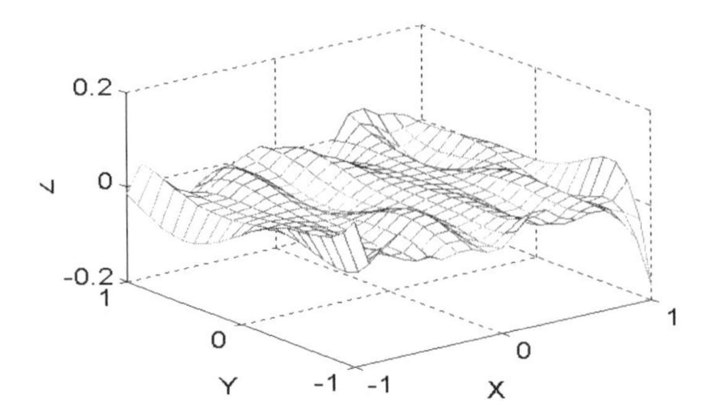

Figure 4.9. Map of the true values calculated with the neural network

From Figure 4.10, it can be seen how the result of the interpolation is satisfied because the error falls within the range [-0.2, 0.2].

The mean square error (MSE) between the true values and those interpolated on all the pixels composing the grid is:

$$MSE = 22.7*10^{-3}$$

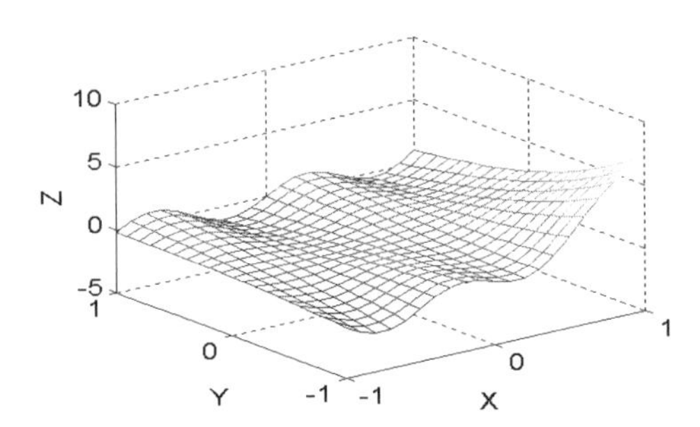

Figure 4.10. Difference map

It can also be observed that the interpolation is not polarized in the sense that the mean of the errors is centered on zero, and in particular has the value of:

$$ME = 4.1*10^{-4}$$

4.5.2 Example of RBFN as a Classifier: Diagnostics Aboard a Railroad Train

This example of using RBFN as a *classifier* describes the study of a *diagnostic system* designed to operate on board a *railroad train* [33]. The purpose of this system is to operate a continuous monitoring of the locomotive's state variables, recorded by a dedicate system of data acquisition.
The sensor network makes various information available to the driver, translating it into a linguistic-type form.
The diagnosis of any fault is thus carried out on line even in those cases of malfunctioning that do not necessarily cause the switching on of the extra-rapid cut-out, or have some difficulty in doing so, with the resultant stopping of the convoy.
The proposed diagnostic system thus foresees three different modes of functioning and the commutation between one of them and another is managed completely automatically:

- *normal operation*, during which the system merely transfers to the visualization system the state variable values_that may be of interest for driving the locomotive;
- a *pre-alarm state* which comes on when the diagnostic system perceives a *non-critical anomaly* in the circuit or functioning (one which does not cause the train to stop); in this case the operator has immediate availability to all the information on the most probable source of the anomaly;
- *state of breakdown*, that occurs as a result of the intervention of one or more *automatic protection devices*; in this case, the diagnostic system carries out the diagnosis and informs the operator of the operations required for starting up again.

In each case, the problem is one of recognizing rapidly the various typologies of faults or anomalous performance, and warning the driver in good time of the most suitable operations needed for starting up again.

It was thus decided to use neural networks and in particular the RBF architecture which is particularly suitable for pattern recognition applications [34].

The variables containing information on the functioning of the locomotive are represented as vectors. The training of the neural network is carried out by use of synthetically obtained fault patterns. A fault vector is made up, as shown in Figure 4.11a, of 100 integer value elements between 0 and 127, including suitably discrete reports both on the state of the locomotive and alarms (digital and analogs). In normal operational conditions, the alarms will all have zero values whereas the state reports will have non-zero values.

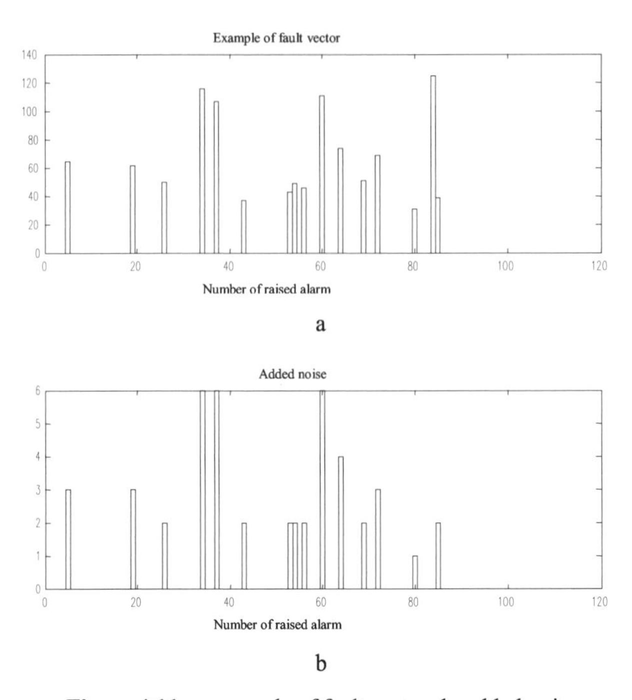

Figure 4.11. a. example of fault vector; b. added noise

Ten different classes of faults were recognized, and the training patterns were generated in conformity with this datum. With the purpose of improving the learning of the network, an zero mean noise with an uniform spectral distribution and with amplitude varying up to ± 20% for the 200 training samples (training set) and up to ± 30% for the 200 test samples (test set), was added. In this way, during the training phase the neural network is strengthened against disturbances that act in the same manner on the real system (errors of measurement, interference, *etc*.). The database thus obtained reflects fairly faithfully the form of the fault vectors which will be used in the final version of the diagnostic system, after making sure that the clustering algorithm has subdivided the input space into a correct number of clusters (that corresponds to finding the most appropriate number of neurons in the hidden layer, which in the case in question was equal to 15). For the mean square error (LMSE) to reach a plateau and stabilize around a value of the order of 10^{-4} for the training set, and of the order of 10^{-3} for the test set (TMSE), 300 cycles were sufficient.

The tests carried out on the network trained in this way afforded, a sure and correct classification of the data in 87% of the cases, while in the remaining 13% of cases there was ambiguity in interpreting the output data; from a more accurate analysis it was however possible to observe that the incorrectly classified patterns were those affected by a level of noise higher than 10%.

To solve cases of ambiguous diagnosis, we will therefore have to address our attention more to the precision of the sensory devices on board rather than to the diagnostic system itself, or that is to say try to integrate, in the fault vector, other information of which a knowledge may be determinant in characterizing any particular fault situation.

Figure 4.12 represents the RBFN model used, and, finally, an example of correct diagnosis is given.

Correct diagnosis:

```
R.B. Units = [ 0 0 0 0 0 0 0 0 0 0 0 0.939 0 0 0 ]
YCALC  = [ 0.0105 0.009, 0.0106, 0.0104 0.0106 0.0104
0.009 0.0105 0.0104 0.9832 ]
YDES   = [ 0 0 0 0 0 0 0 0 0 1 ]
Diagnoses:              DEFECT #10
Diagnostic certainty:... (YMAX/YMIN)........    108.8
Spectral purity:... (YMAX/YMED):      9.14
```

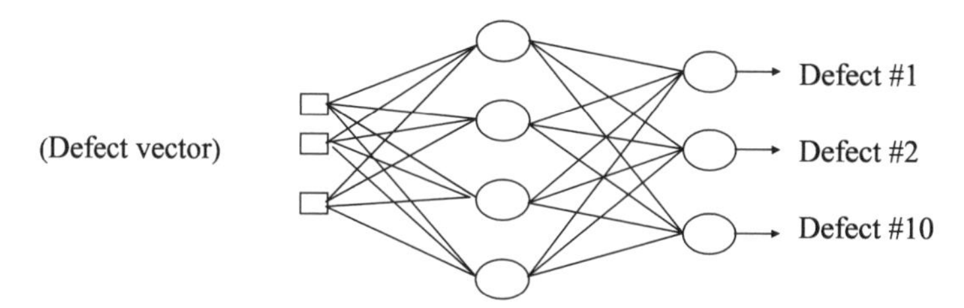

Figure 4.12. Radial Basis Function Network model

Although the problems of classification are those for which RBFNs are most frequently employed, it is still possible to use them also for solving problems of approximation and interpolation of non-linear functions.

To conclude this chapter, it seems appropriate to remark that both the learning and the optimal neural network topology can be considerably improved by use of evolutionist optimization algorithms, which will be described in Chapter 6. In fact, with regard to learning, the genetic algorithms allow us to avoid the local minima of the standard error function and thus supply a better configuration of the weights. The choice of the optimal topology of a neural network may furthermore be easily codified as a problem of optima that can be solved with genetic algorithms, as is done, for example, in [40].

4.6 References

1. Lippmann RP. An Introduction to Computing with Neural Nets. IEEE Acoustict Speach and Signal Processing Magazine. 1987; 4: 2: 4-22
2. Hush DR, Horn BG. Processing in Supervised Neural Networks. IEEE Signal Processing Magazine. January 1993
3. Zurada J M. Introduction to Artificial Neural Systems. West Publ. Comp. St. Paul, MN,1992

4. Chu SR, Shoureshi R, Tenorio M. Neural Networks for System Identification. IEEE Control System Magazine. April 1990; 31-34
5. Fortuna L, Graziani S, Muscato G, Nunnari G. Auto-associative Multi-layered Neural Networks for the Classification of Seismic Signals. IJCNN, Baltimore, June 1992
6. Carelli R, Camacho EF, Patino D. A Neural Network Based Feedforward Adaptative Controller for Robots. IEEE Trans. on Systems, Man and Cybernetics. 1995; 25: 9
7. Rosenblatt F. The Perceptron: A Probabilistic Model for Information Storage and Organization in the Brain. Psycological Review. 1958; 65: 386-408
8. Masters T. Practical Neural Network Recipes in C++. Academic Press, Inc., USA, 1993
9. Cybenko G. Approximation by Superposition of a Sigmoidal Function', Mathematics of Control, Signals and Systems. 1989; 2: 4: 303-314
10. Funahashi K. On the Approximate Realization of Continuous Mappings by Neural Networks. Neural Networks. 1989; 2: 183-192
11. Chen S, Billings SA, Grant PM. Non-linear System Identification Using Neural Networks. Int. J. Control. 1990; 51: 60: 1191-1214
12. Narendra KS, Parthasarathy K. Identification and Control of Dynamical Systems Using Neural Networks. IEEE Trans. on Neural Network. 1990; 1: 1: 4-21
13. Rumelhart DE, McClelland JL. Parallel Distributed Processing: Explorations in the Microstructure of Cognition: Foundations. vol.1, MIT Press, Cambridge, MA, USA, 1986
14. Arena P, Fortuna L, Graziani S, Muscato G. Simulation of Multilayer Perceptron with Automatic Tuning of Learning Parameters. European Simulation Conference ESM Copenaghen, 1991
15. Arena P, Fortuna L, Graziani S, Nunnari G. A Monitoring Approach for the Design of Multi-layer Neural Networks. Proc. of the Third International Congress on Condition Monitoring and Diagnostic Engineering Management. UK, 1991
16. Karnin ED. A Simple Procedure for Pruning Back-propagation Trained Neural Networks. IEEE Trans. on Neural Networks. 1990; 1: 2
17. Weigend AS, Rumelhart DE, Huberman BA. Generalization by Weight-Elimination with Application to Forecasting. in Touretzky, D., editor Neural Information Processing Systems. Denver, Morgan Kaufmann 1988; 1
18. Arena P, Fortuna L, Graziani S, Muscato G. A Real-Time Implementation of a Multilayer Perceptron with Automatic Tuning of Learning Parameter. Algorithms and Architectures for Real-Time Control, IFAC Workshops Series. 1992; 4
19. Karnin CD. A Simple Procedure for Pruning Back-propagation Trained Neural Networks. IEEE Trans. on Neural Networks. 1990; 1: 2: 239-242
20. Sietsma J, Dow RJ. Neural Networks Pruning-Why and How. Proc. IEEE Int.Conf.on Neural Networks, San Diego, California, vol. I. 1988; 325-333
21. Carpenter GA, Grossberg S. A Massively Parallel Architecture for a Self-organising Neural Pattern Recognition Machine. Computer Vision, Graphics and Image Proc. 1987; 37: 54-115
22. Kohonen T. Self organized Formation of Topologically Correct Feature Maps. Biological Cybernetics Reprinted in Anderson and Rosenfeld. 1988; 43: 59-69
23. Chen S, Cowan CFN, Grant PM. Orthogonal Least Squares Learning Algorithm for Radial Basis Function Networks. IEEE Trans. on Neural Networks. 1991; 2: 2
24. Botros SM, Atkeson CG. Generalization Properties of Radial Basis Function. in D. S. Touretzky, editor, Advances in Neural Information Processing System III. Morgan Kaufmann, San Mateo, CA 1991; 707-713

25. Arena P, Fortuna L, Gallo A, Graziani S, Muscato G. Induction Motor Modelling Using Multi-Layer Perceptrons. IEICE Trans. on Fundamental of Electronics, Communication and Computer Sciences. 1993; E-76-A: 5
26. Antsaklis PJ. Special Issue on Neural Networks for Control System. IEEE Control Syst. 1990; 10: 3: 3-87
27. Antsaklis PJ. Special Issue on Neural Networks for Control System. IEEE Control Syst. 1992; 12: 2: 8-57
28. Fortuna L, Muscato G, Nunnari G, Occhipinti L. Neural Modeling and Fuzzy Control: An Application to Control the Temperature in a Thermal Process. Proc. IEEE int. Conf. Fuzzy Syst., vol. II. San Francisco 1993; 1327-1333
29. Hecht-Nielsen. Neurocomputing. Addison Wesley, Reading, MA, 1990.
30. Kohonen TK. Self-Organization and Associative Memory. 3rd ed. Springer-Verlag, New York, 1989
31. Kosko B. Neural Networks for Signal Processing. Prentice-Hall Englewood Cliffs, NJ, 1992
32. McClelland TL, Rumelhart DE. Parallel Distributed Processing. MIT Press Cambridge, MA and the PDP Research Group, 1986
33. Gallo A, Nunnari G, Muscato G, Occhipinti L. Diagnostica a Bordo di Rotabili Ferroviari, Approccio Basato Sulle Reti Neurali. Automazione e Strumentazione. Maggio 1995; 125-130
34. Holler M. et al. A High Performance Adaptative Classifier Using Radial Basis Function. Gomac, 1992
35. Hsia TC. System Identification: Least-Squares Methods. D. C. Heath and Company, 1977
36. Ljung L. System Identification: Theory for the User. Prentice Hall, Upper Saddle River, NJ, 1987
37. Cybenko G. Approximation by Superposition of a Sigmoidal Function. Math of Control Signals and Systems, Springer, New York, 1989: 2: 303-314
38. Freeman JA, Skapura DM. Neural Networks - Algorithms, Applications and Programming Techniques. Addison-Wesley, 1992
39. Rumelhart DE, Hinton GE, Williams RJ. Learning Internal Representation by Error Propagation. In: Parallel Distributed Processing. MIT Press, Cambridge, MA, 1986; 1: 18-362
40. Arena P, Caponetto R, Fortuna L, Xibilia MG. MLP Optimal Topology via Genetic Algorithms. Proc. ANNGA '93. Innsbruck 1993
41. Arena P, Caponetto R, Fortuna L, Muscato G, Xibilia MG. Quaternionic Multi-Layer Perceptron for Chaotic Time Series Prediction. IEICE Trans. on Fundamentals. 1996; E-79-A: 1647-1657
42. Arena P, Fortuna L, Muscato G, Xibilia MG. Neural Networks in Multidimensional Domains. Lecture Notes in Control and Information Science. Springer-Verlag, Vol. 234, 1998
43. Arena P, Fortuna L, Gallo A, Nunnari G, Xibilia MG. Air Pollution Prediction via Neural Networks. Proc. 7th IMACS-IFAC Symp. LSS'95, London. July 1995

5. Neural Networks for Modeling and Controlling Dynamical Systems

5.1 Identification of Non-linear Systems

The set of properties characterizing a given system assumes the name of *model* of that system. In the literature, models of various types have been defined [1]:

- mental or *intuitive models*, listed by a set of logical inferences residing in the human mind and not expressed analytically;
- *graphic models* in which the properties of the system are represented by graphs or tables;
- *mathematical models* in which the relations connecting the variables of a system one to another are expressed by means of differential equations.

Mathematical models are those which for obvious reasons are the most interesting. In order to obtain a mathematical description of a system, the *direct approach* is that of investigating the relationships and physical laws that regulate the interactions between the variables inside the system and that generate the output signals. The strategy described is often called *modeling*.

Frequently this type of direct approach is not possible for reasons that concern the knowledge one has of the system, that at times can be incomplete, and because the properties of the system vary in function of some not easily identifiable conditions. Moreover, the modeling can in certain cases be very difficult and lead to models that are too complex to be used, or conversely too poor, due to the simplifications they contain. Thus recourse must be had to strategies that make direct use of the measurements of those signals produced by the system for constructing the model. These strategies are known as *identification* ones [10]. It should be noted that the measurement of the input/output signals of a system for constructing a model depend on the particular physical context, *i.e.*, on the laws generating the observed signals.

The approach to the problem of determining a model which uses the measured data exclusively is called *black box identification*: this has proved an indispensable tool for solving various practical problems of great interest.

For identification, a further distinction can also be made between *parametric identification* and *non-parametric,* or *spectral identification,* according to whether the aim is to determine a vector of parameters or to determine the frequency response.

To construct a model starting from the input/output data, apart from the dimensions, the following points should be considered:

- a set of candidate models;
- a rule that will allow the best model from among the various candidates to be chosen on the basis of the dimensions.

The latter point is implicit in the identification model used. The choice of model is typically suggested by the way in which the model itself is able to reproduce the measured data, in that the vector of the parameters forming the model exemplifies the hidden manner in which the process elaborates the input data. Once the parameters have been obtained, the model can be used for estimating or predicting future data on the basis of those already known. In this respect, it is clear that any studies on dynamic system identification are inevitably involved in, and sometimes overlap, the methods for processing signals and analyzing time series. The goodness of the model is thus assessed on the basis of the error between the output of the model and that of the system in question at the same moment in time. Ideally, this error should be zero at any moment, but in practice it is sufficient to let its trend obey the appropriate statistical characteristics, which is to say that it should be as close as possible to an uncorrelated signal with zero mean, *i.e.*, to a white noise.

In the field of *linear systems*, much of the development of identification techniques and many of the other strategies employed are based on the choice of an appropriate family of auto regressive models, the so-called ARMAX (Auto-Regressive, Moving Average with eXogenous inputs) models. It is thus hypothesized that a discrete linear system can be described by a linear combination of a certain number of samples of output, input, and of the error, modulated by parameters that must be determined by means of the identification procedure. The ARMAX model is of this type:

$$y(k) = a_o + \sum_{i=1}^{n_y} a_i y(k-i) + \sum_{i=1}^{n_u} b_i u(k-i) \tag{5.1}$$

where y and u, are respectively the output and input signals, while a_i and b_i represent weight coefficients. Whereas the techniques for identifying linear systems are well established and fully experimented with a great number of applications, numerous difficulties arise from the complexity of the phenomena and behaviors which a non-linear system is able to generate. To this should be added the difficulty of deriving a mathematical representation which is useful for describing vast classes of non-linear systems. All this leads inevitably to the use of a great number of parameters for characterizing even very simple non-linear systems. Traditional identification of non-linear systems can make use various approaches [2].

When the structure of the system to be identified is not known *a priori* and a black box approach is taken to the modeling process, the approach generally utilized is that of characterizing the system output at a given moment such as the processing of an expansion of the input-output dimensions by means of a non-linear function F (NARMAX model):

$$y(k) = F[y(k-1),..., y(k-n_y),u(k-1),..., u(k-n_u)] \tag{5.2}$$

Identification by NARMAX structures is nothing more than a generalization of the ARMAX approach in the case of linear systems [3].

It can be inferred that a NARMAX model may represent a vast class of systems. Those with distributed parameters and those that do not allow linearization around an operating point are, however, excluded. The written equation refers to a system having one input and one output only, although the same considerations are valid in multivariable cases. Moreover, it is well known [4] that a NARMAX model may represent a vast class of non-linear systems, including the particular cases of bi-linear representations and many others.

The classical approach followed for deriving NARMAX models of non-linear systems makes use of the well-known *Stone-Weiererstrass theorem* [5] which states that the set of all the real polynomial functions is *dense* in the space of continuous functions and defined on a compact set. Therefore the procedure consists in expanding relation 5.2 into a series of polynomial fixed-order functions and in applying appropriate iterative algorithms such as the extended least squares recursive algorithm [6].

A basic problem in non-linear identification is the choice of *model structure*, even before making a complete estimate of the coefficients. A model in which the polynomial order is high requires an excessive number of parameters. Thus it is better to carry out careful analyses regarding the contribution of each output term. Independently of the identification algorithm considered, particular attention should be addressed to validating the model.

In non-linear systems, the validation as a function of the prediction error is translated into a condition of unpredictability of all the previous linear and non-linear input and output combinations. This condition may be verified by analyzing the *residue/output* and *residue/input* correlations of the system.

5.2 Identification by Means of Neural Networks

In the previous section, the problem of identifying non-linear systems was considered. If one takes the last class of non-linear systems described, *i.e.*, the NARMAX models, it can be seen how a NARMAX model processes a certain number of measured samples in a given temporal window by a non-linear function, F, in such a way to obtain the estimated output at a certain subsequent moment. Sometimes the function F is known *a priori* and the identification procedure merely involves the determination of a number of fixed parameters. In other cases, however, very little is known of the system and the only sources available are the input/output samples.

The approach using a *polynomial expansion* of the NARMAX structure is a valid method, guaranteed furthermore by the presence of theoretical results, to which reference has been made, concerning the approximation capacity by means of polynomials.

An alternative approach is that of utilizing *neural networks* as models for identifying non-linear systems. The *weights* of the network constitute the *parameters* to be identified, and the *learning algorithm*, by minimizing the square error between the output of the system and that of the model, assumes the role of an *identification algorithm*.

With regard to approximation capacity, the theoretical results given in the previous section are valid. Therefore, in the case of identifying a non-linear system by means of a neural approach, the network inputs correspond to the input/output samples of the system itself with reference to the previous moments of time, whereas the unknown function F, instead of being developed in polynomials, is iteratively approximated by the learning algorithm as a linear combination of sigmoidal functions.

The conditions for a NARMAX model to exist, although they can be easily verified by a large number of real systems, are locally valid, that is to say in a region around an equilibrium point of the system itself. Furthermore, it is assumed, from the standpoint of input/output, that the function describing the dynamics of the system is continuous and can be differentiated in that region [3].

From this point of view, and in the light of the theorems reported in the previous chapter, it can be stated that a neural network implements a generalized NARMAX model of a non-linear system, in the sense that an approximation by a sequence of sigmoid functions is valid for piecewise functions and in vast operational areas [13].

Therefore, if a system can be represented by different NARMAX models in different working areas separated by points of discontinuity, there is one neural network only that can interpolate the dynamic behavior of the excitation region. The steps to be followed to obtain a sufficiently accurate model of a system were outlined at the beginning of this chapter; however, the use of neural structures for modeling requires further study and in particular with regard to the following points [7]:

- the representation of the problem, *i.e.*, determination of the variables (inputs, outputs, states) which must be taken into account when constructing the model, and especially the way in which such information should be codified. This choice influences in a determining way the result of the modeling in that, in a certain sense, the network operates a certain codification on the input representation. In such a framework, no theoretical formulations exist, and thus much depends on experience and heuristics;
- the choice of the classes of models; in our case, the choice of the number of inputs for the network the order of the model and the optimal number of parameters, which corresponds to the number of neurons required by the network for minimizing the behavioral indices.

With regard to the number of inputs for the network, the use of appropriate tests (which will be dealt with below) extrapolated from linear systems theory may well lead to the correct choice being made:

- validity criteria: as with all identification criteria, also that for neural networks requires validation methods that impose conditions on the method's reliability (and these should not be based solely on minimizing the error in the learning and generalization phases).

- *a priori* knowledge: certainly, what is known about a system is of valid help in constructing a model by means of classical approaches, and any information will be useful for choosing the class of model, the initial values of the parameters, and even the bounds of the various parameters.

When identifying a system with neural networks, it is often almost impossible to introduce the knowledge one has of the system, because the model one obtains is a distributed representation of the system. It is thus impractical (admitting that it is possible) to identify that part of the connections matrix which identifies a given portion of the model's structure. Undoubtedly, however, it is possible to introduce any knowledge of the system in terms of a suitable codification of the input signals and of the number of input units of the network, provided one knows for example the approximate order of the system.

5.3 Effects of Disturbance and Uncertainty

In linear systems, the presence of a disturbance can be modeled as an additive term on the output. The same cannot be said of a linear system which, generally, has to deal with terms relative to linear and non-linear combinations of the disturbance, both with the output and input of the system.

The network input patterns are always presumed to be subject to disturbance, and indeed they are often voluntarily *dirtied* in order to *stimulate* the capacity, which is typical of neural networks, of operating a filtering of the uncorrelated component present in the input signals; this is all to the advantage of the abstraction of a rule, which in our case coincides with the map describing the system's dynamic behavior.

Nor should the uncertainty inherent in the parameter vector be forgotten: since this vector only ever represents an estimate of the set of real parameters, the phenomenon is therefore called *parametric uncertainty*.

The effects of a disturbance which is not efficaciously filtered, or those of a relevant uncertainty in the parameters, can give rise to heavily polarized estimates, and thus to inadequate models.

In the case of an estimate made by a neural approach, the polarizing of the model does not necessarily imply a bad fit of the signals utilized for learning. Rather, since the learning algorithm can minimize the square error at the end of the learning phase, it will give a good estimate regarding the window of data utilized for the learning.

However, this condition does not necessarily indicate a goodness of the network for what concerns a definition of a good model for the system to which the data refer. In fact, the model obtained can be described by a function which interpolates the window of data supplied during learning, but which has little to do with the peculiar characteristics of the system which has generated those data. That may be seen in relation to the previously described overfitting phenomenon.

It can therefore happen that the error in the testing phase, once it reaches a minimum assumes increasingly greater values as the learning cycles increase. Under these conditions, as was concluded in the previous chapter, only a slight amount of information is found in the learning patterns, or else an inadequate number of parameters is being employed, and thus a wrong topology. *vice versa*, an adequate number of parameters can lead to a polarized estimate if the order of the system has been underestimated, or else if at the input to the network a sufficient number of input/output sample delays has not been given.

This aspect is directly linked to the problem of structural uncertainty in modeling the system.

Moreover, it can be shown that a neural network does not generate delays, *i.e.*, it does not display system input/output components different than those at the input nodes.

In fact, the neural network carries out a static mapping and yields a *static expansion* of the information supplied at the input points [8].

These considerations impose a procedure of verification and validity for the model; this consists essentially in an examination of the statistical characteristics of the error between the model and the system, and in particular of *self and cross-correlation functions* of the error with respect to the inputs and outputs of the network.

The procedure derives from a generalization of the theory of parameter estimates in linear systems with the least square method. In such cases, in fact, the well-known pseudoinverse logarithm yields:

$$\hat{\vartheta} = (\mathbf{\Phi}^{T}\mathbf{\Phi})^{-1}\mathbf{\Phi}^{T}y \qquad (5.3)$$

where:

$y = \mathbf{\Phi}\theta + \varepsilon$ is the system output;
θ is the parameters vector;
Φ è is the measurement matrix;
ε is the vector of the residuals.

In these conditions, in order to obtain a non-polarized estimate, that is to say one in which on average the estimated parameter vector tends to a true value, one must have:

$$E(\hat{\vartheta}) - \vartheta = 0$$

which is to say:

$$E(\mathbf{\Phi}^{T}\varepsilon) = 0.$$

That means that the input-output measurements which form the parameters of the matrix $\mathbf{\Phi}$ and the value of the residuals ε, must be uncorrelated.

Generally, for a non-linear model, a validity procedure takes into account the correlation between the residuals and all the linear and non-linear combinations of the input/output samples.

Billings and Voon [9] showed how that is translated into the following conditions:

$$\Phi_{\varepsilon\varepsilon}(\tau) = E[\varepsilon(t-\tau)\varepsilon(t)] = \delta(\tau)$$

$$\Phi_{u\varepsilon}(\tau) = E[u(t-\tau)\varepsilon(t)] = 0, \forall\tau$$

It can be noted that the relations reported have been obtained from the analytical non-linear systems. The neural networks representing the NARMAX models are able to approximate vaster classes of systems, which is one reason why these correlation tests do not constitute a necessary condition for guaranteeing non-polarized estimates.

Although, due to mathematical difficulties, no thorough theoretical analysis of the problem has yet been made, the correlation tests offer valid help in identifying a good neural model. It should also be emphasized that these tests are useful for identifying the lower limit for the number of inputs in the network, and they can thus be classified as tools for helping the techniques of *growing*; of course, once the order of the system is known, this analysis can be avoided. It should also be noted that by increasing the number of network inputs, the characteristics in the generalization phase can be improved, whenever a greater number of parameters, and not of sigmoids, is required.

An important phase for deriving a model is the collection of measurements regarding the system to be identified. It is clear that this phase depends on the ability or inability of the system to be measured *ad hoc.* When this occurs, the strategy typically adopted is that of exciting the system with an input signal from a extremely rich harmonic content. The amplitude and bands of the excitation signal depend on the knowledge one has of the dynamic characteristics of the system.

5.4 Example of Identification by Means of a Neural Network

This example briefly describes the identification of the model of a solar energy system for heating a house [10], obtained by using a neural network of the perceptron type with one hidden layer. This system is represented schematically in Figure 5.1.

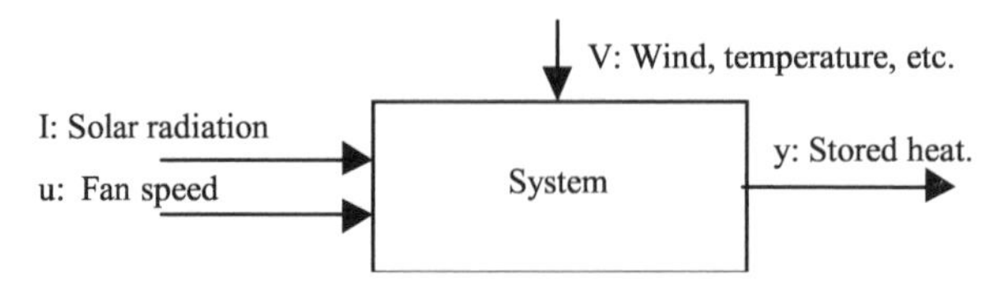

Figure 5.1. Model of a heating system

The system operates in the following way. The sun heats the air in the solar panel, which is formed of transparent tubes. The warm air is then conveyed by a fan to a heat tank made available for heating the house. The (time-discrete) model that has to be identified is described by Equation 5.4:

$$y(t) = (1-d_1)y(t-1) + (1-d_3)\frac{y(t-1)\,u(t-1)}{u(t-2)} + (d_3-1)(1+d_1)\frac{y(t-2)\,u(t-1)}{u(t-2)}$$
$$+ d_0 d_2 u(t-1)I(t-2) - d_0 u(t-1)y(t-1) + d_0(1+d_1)u(t-1)y(t-2) \tag{5.4}$$

It can be noted that the output y at the instant t depends not only on U and I, calculated at previous instants, but also on the same value calculated at the instants t-1 and t-2. The model is therefore of a regressive non-linear type. The neural network utilized for the identification has 5 neurons in the input layer (corresponding, respectively, to $y(t$-$1)$, $y(t$-$2)$, $u(t$-$1)$, $u(t$-$2)$, and $I(t$-$2)$, 16 neurons in the hidden layer (the value chosen for subsequent attempts), and only one output neuron ($y(t)$).

The patterns utilized for learning (2000) and for the test (400) were obtained from (5.4), assigning the input variables, which have the trend shown in Figure 5.2, the coefficients d_i, and obtaining the corresponding values for the output shown in Figure 5.3. The learning cycles were 5000 with a mean square error for the test (TMSE) equal to ca 10^{-2}. White noise, equal to 5% of the signal in question, was added both to the input values and to the output one. The values assigned to the coefficients d_i were: $d_0 = 0.3$, $d_1 = 0.6$, $d_2 = 2$, $d_3 = 1.3$. It should be pointed out that normally in black-box-type identification procedures, like the one described, which are obtained with neural procedures, the model is not known *a priori* and is obtained on the basis of measuring the input and corresponding output signals. To identify the system means actually to reproduce the trend of (5.4) and thence to identify the above-mentioned coefficients not known *a priori*.

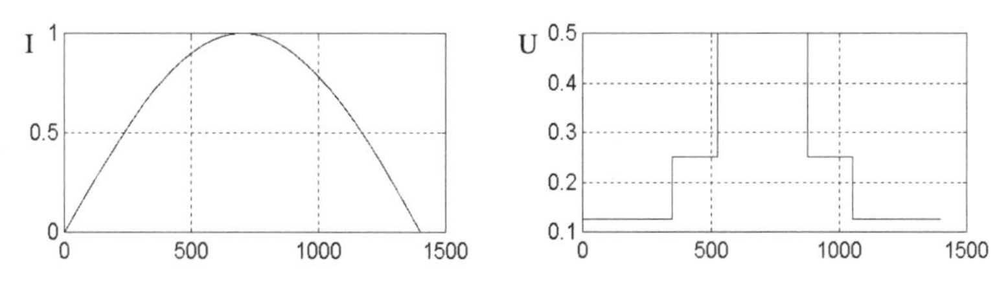

Figure 5.2. Trend of the variables I and U over time

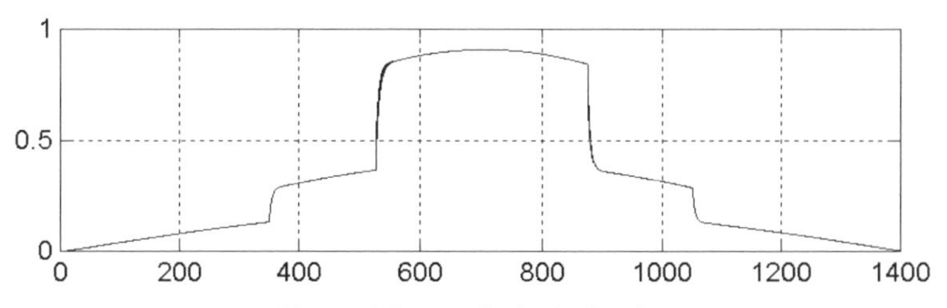

Figure 5.3. Trend of calculated *Y*

To test the validity of the model obtained, the following procedure was carried out. For the signals *I* and *U*, two randomly trending signals were taken, variable within appropriate ranges. Then the corresponding trend for the output signal was obtained, both for Function 5.4 (the *true* signal) and for the model identified. The two signals are represented in Figure 5.4 with respect to the first 100 samples obtained in the time (for reasons of clarity of representation). The trend of the true signal is represented by the continuous line and that of the model identified with a broken line. Finally, Figure 5.5 shows the difference between the two signals.

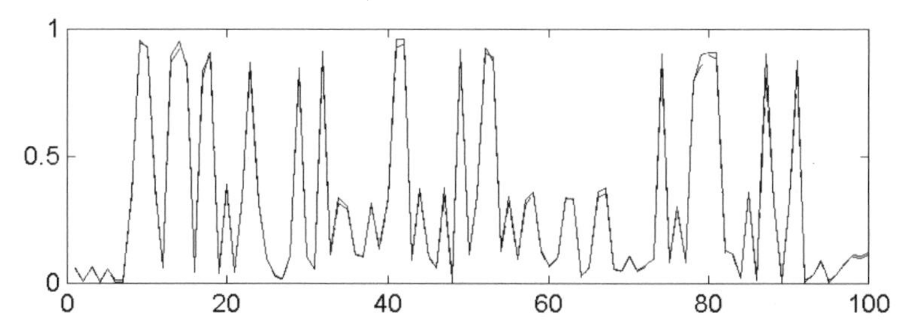

Figure 5.4. Overlap of the signal calculated and the model output

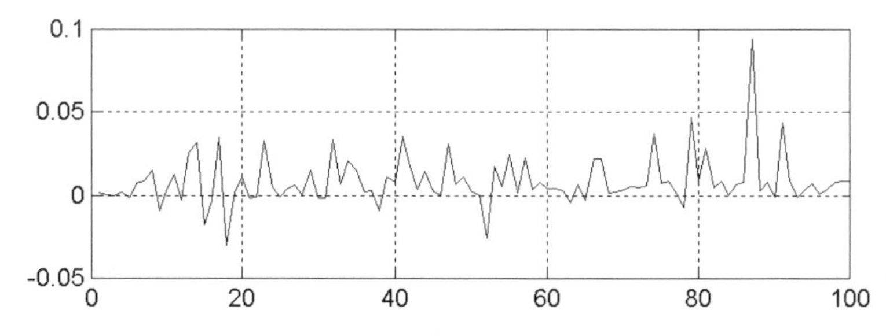

Figure 5.5. Difference in signal

From Figure 5.5 it can be seen how the model identified is relatively satisfactory, the percent maximum distance between the two signals not being significant.

5.5 Neural Control

The use of neural networks has led to increased interest also in the field of control, and in particular in process control, robotics, manufacturing industries, and in applications for space [11], [14-17].

It should be recalled here that the purpose of the controller is that of supplying appropriate inputs for a system (process to be controlled) so that the system exhibits the desired behavior.

In the specific case in which the control action is performed by a neural network, we refer to a *neural controller.*

In order to understand the action of a neural controller, it should be recalled that the control of a process can generally be obtained according to the open-loop control or the feedback control strategy.

It is well known that the open-loop control is bound to failure in all those cases where the transfer function of the process varies in the course of the operation or is not known with adequate precision.

It can also be observed that open-loop control requires a knowledge of the inverse dynamics of the plant, and that may be obtained by means of an identification phase of the process to be controlled; in the neural case, such an operation is known as the *learning control* phase.

The identification of a dynamic, either direct or inverse, can be obtained by utilizing multilayer perceptron-type neural networks, which are trained by means of the well-known back-propagation technique, as discussed previously. The basic configurations for neural identification of a system are shown in Figure 5.6 below.

In the scheme represented in Figure 5.6a, the neural network receives the same input, $\boldsymbol{x}$, as the plant, whereas it is trained in such way that its output, $\boldsymbol{o}$, is equal to that of the plant, $\boldsymbol{d}$. It is thus evident that in such cases the purpose is one of identifying the direct dynamics.

Instead, the inverse dynamics of the plant can be identified by means of the scheme given in Figure 5.6b; in that case, the norm to be minimized is the difference between the input vector $\boldsymbol{x}$ (the reference of the control system) and the vector $\boldsymbol{o}$.

However, it should be pointed out that when the problem of identifying inverse dynamics arises, some other problems may arise in those cases where the dynamics is not unequivocally defined.

During the identification phase, for the multilayer perceptron, a configuration of the type depicted in Figure 5.6c might be used; this is characterized by the fact that in the input layer, in addition to the values of $u(k)$, $u(k\text{-}1)$, *etc.*, also the output values $y(k\text{-}1)$, $y(k\text{-}2)$,... $y(k\text{-}p)$ are present and are measured in the instants preceding k.

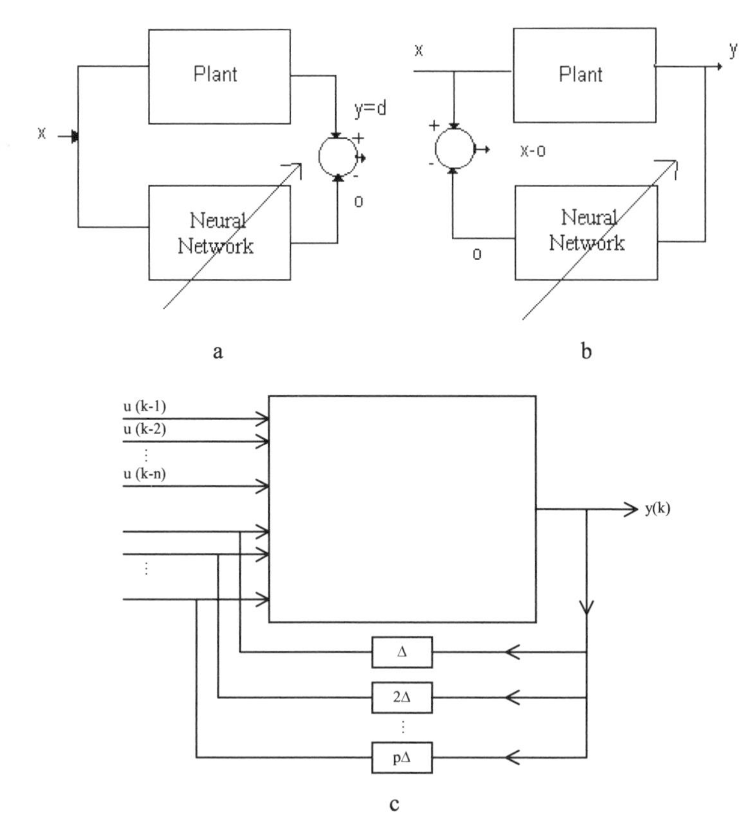

Figure 5.6. a. identification scheme for neural control, direct dynamics identification; b. identification scheme for neural control, inverse dynamics identification; c. identification of a dynamic system by means of a neural network

5.6 Feed-forward Control with Identification of the Inverse Dynamics

Figure 5.7 shows a feed-forward control scheme set up utilizing a neural network *A* which is trained on line to identify the inverse dynamics of the plant. The input of the control system is ***d***, which represents the response required from the plant, ***x*** should be the input signal devoted to control the real plant. The real output of the plant is instead indicated by ***y***.

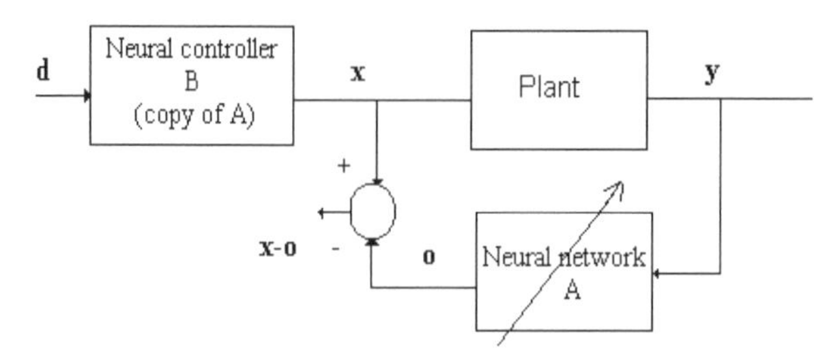

Figure 5.7. Feed-forward control with learning of the inverse dynamics

The error used to train the neural network *A* is the difference between the outputs of the neural networks *B* and *A*, respectively. It is easily understood that in those hypotheses in which the inverse dynamics of the plant is univocally defined, the system will evolve towards a condition in which the difference *x-o* is reduced to zero, that is to say where ***y*** exactly follows ***d***. This control architecture, known as indirect learning architecture, offers undoubted advantages: the neural network indicated with A can be trained on-line, whereas its replicate *B* carries out the role of controller. Furthermore, since the inputs to the control system are the desired outputs, the training work can be done in the considered domain (specific training), thus allowing better performances to be achieved throughout the system. Finally, it is observed that the control scheme in question can function in an adaptive way, and therefore with a plant whose dynamics can be time-variant.

Another way to implement a neural control system which will have characteristics similar to those described above, but will utilize one neural network only, instead of the two required by the scheme shown in Figure 5.7, is that given in Figure 5.8.

The reader will be able to verify without difficulty, by reasoning as outlined above, that if the neural controller is trained so as to minimize the norm of the vector ***d- y*** the target of the control is reached.

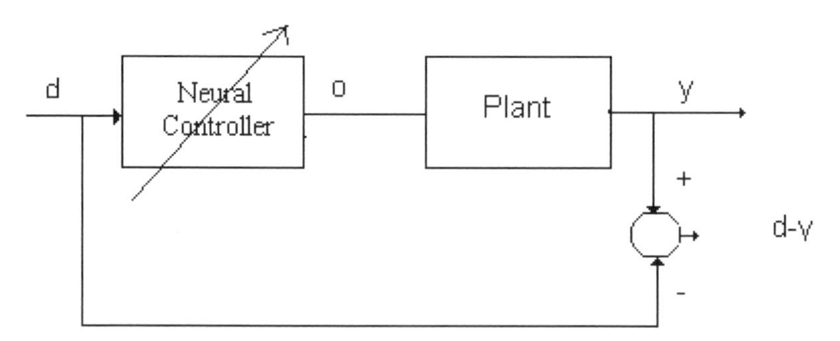

Figure 5.8. Scheme of on-line neural control with skilled learning

5.7 The CMAC Control Scheme

This section will describe the cerebellar model articulation controller (CMAC) control scheme shown in Figure 5.9.

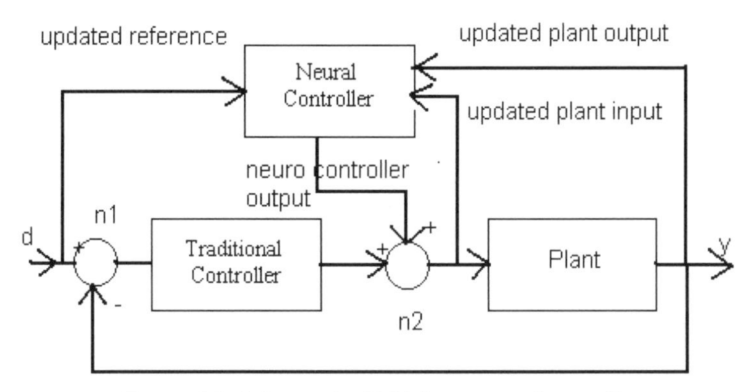

Figure 5.9. Scheme of a CMAC-type neural controller

From the scheme it can be seen that *two controllers* are present: one labeled *traditional controller* and the other *neuro-controller*. It is also observed that the traditional controller is inserted inside a classical feedback control loop. The mechanism inside the system is the following: in the initial phase, the weights of the neural networks implementing the neural controller are made equal to zero, and thus the output of this controller, which is summed at the node *n2*, is nil.

Instead, in this phase it is the traditional controller that supplies the necessary output for controlling the plant; therefore, hypothesizing that the feedback control loop is asymptotically stable and that the loop gain is sufficiently high, **y** will be found approximately equal to ***d***. However, at each control phase, the signal produced by the traditional controller, which we can denote with $\boldsymbol{x}(k)$, where k is the time-discrete instant, is utilized for constructing a learning rule of the type $w(k+1)=w(k)+\alpha(x(k)-w(k))$.

In this expression for adapting weights, the constant α is a small positive constant (learning constant). During the learning phase, the control signal produced by the traditional controller can be summed in the n2 node with that of the neuro-controller.

This procedure produces a control signal that will grow from zero until it gradually assumes complete control of the plant. Instead, since the traditional controller is excited by the error signal produced at the output of the summing node *n1*, which will then gradually drop to zero, it will necessarily be excluded from the control action.

An immediate consequence of the fact that the feed-forward-type of neuro-controller shown in Figure 5.9 assumes control of the plant is an increase in speed of response of the whole control system. In the presence of noise or variations in plant parameters, however, the traditional controller may, even if only partially, re-assume control in order to compensate for uncertainties while waiting for the self-learning capacity of the neuro-controller to make it able to adapt to any variations that might have arisen.

A neural control scheme whose architecture is similar to that shown in Figure 5.9, and which has been proposed in the literature [12] for applications in the field of robotics, is shown in Figure 5.10 below.

The modeling of the manipulator dynamics is carried out on-line. At each cycle, the neural network weights are up-dated on-line so as to minimize the feedback error. Namely, if one indicates with $\boldsymbol{o}^{(k)}$ the output of the neural network, and with $\boldsymbol{u}^{(k)}$ that of the traditional controller at step k, the new target of the neural network is fixed as equal to the sum of $\boldsymbol{o}^{(k)} + \boldsymbol{u}^{(k)}$. Thus, the network is continually learning from the feedback error received from the traditional controller; hence the neural network learns the output value to be produced in order to impose the desired output on the plant, and in this way the feedback error is reduced to zero.

The reader will thus be able to recognize without difficulty that, once this condition is reached, the control system will behave like a feed-forward system.

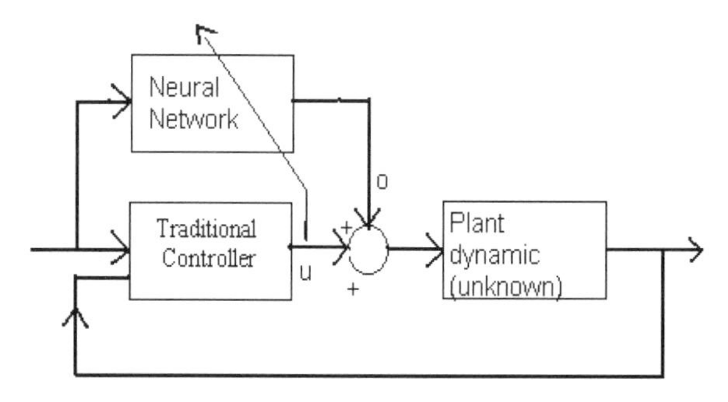

Figure 5.10. Scheme of a neural control experimented in robotics applications

While trajectories are being performed in real time, the network weights are being continuously up-dated by means of the back-propagation algorithm to a frequency that is compatible with the on-line operation of the manipulator.

To summarize the characteristics of the neural control schemes so far considered, we may use that of Figure 5.11.

In this scheme, the presence both of a feed-forward- and of a feedback-type controller can be seen; furthermore, the presence of a block labeled *observer* is noted, whose role is that of identifying the on-line system model. The information supplied by the observer may thus be utilized for adapting the parameters both of the feed-forward and feedback controller. It is evident that both the observer and the feed-forward controller can have a neural structure, like that depicted in Figure 5.7. Other neural control schemes present in the literature are shown in Figure 5.12a and b.

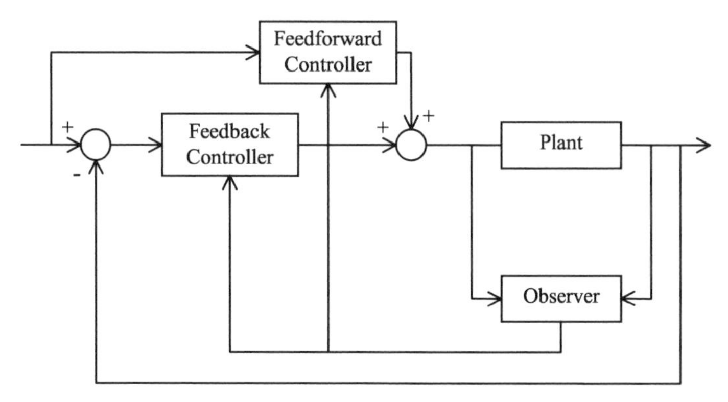

Figure 5.11. A general scheme of neural control

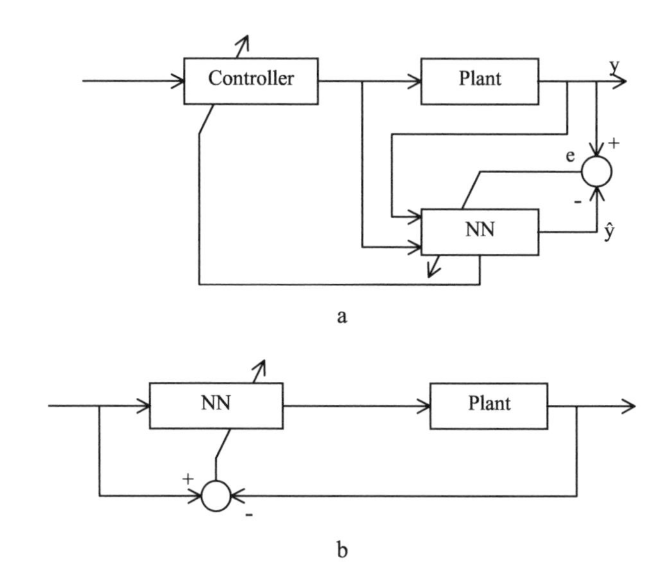

Figure 5.12. a. and b. neural control schemes

The reader will have no difficulty in recognizing in these schemes the ones described above and the relative functioning. To conclude these brief notes on neural-type controllers, in which we have merely illustrated the basic concepts rather than studied in any great depth the applicational and technological aspects, it can be observed that the numerous studies that have appeared in the literature concerning this type of controller generally indicate their superiority over the traditional ones. At the same time, it should not be forgotten that the results reported often refer to simulations and that the applicational type of problems are often not mentioned, whereas they can instead be very relevant. In Chapter 13, the reader will find an application of neural control for the case of an industrial manipulator.

5.8 References

1. Bittanti S. Identificazione Parametrica. CLUP
2. Billings SA. Identification of Nonlinear Systems – A Survey. Proc. IEE Vol.127, Part D, N.6. 1980
3. Leontaritis J, Billings SA. Input-Output Parametric Models for Non-linear Systems. Part I: Deterministic Non-linear Systems, Int. J. Control. 1985; 41: 2303-2328
4. Chen S, Billings SA. Representations of Non-linear Systems: The NARMAX Model. Int. J. Control. 1989; 43: 5: 1013-1032
5. Rudin W. Principles of Mathematical Analysis. McGraw-Hill, New York, 1964
6. Billings S A, Leontaritis J. Parameter estimation of non-linear systems. Proc. 6th IFAC Conference on Identification and Systems Parameter Estimation, Washington, 1982: 427-432
7. Miller W T, Sutton RS, Werbos PJ. Neural Networks for Control. The MIT Press, Cambridge, MA, 1990
8. Billings SA, Jamaluddin HB, Chen SA. Properties of Neural Networks with Application to Modeling Non-linear Dynamical Systems. Int. J. Control. 1992; 55: 1: 193-224
9. Billings SA. Voon WSF. Correlation Based Model Validity Tests for Non-linear Models. Int. J. Control. 1986; 44: 1: 235-244
10. Ljung L. System Identification: Theory for the User. Prentice Hall, Upper Saddle River, NJ, 1993
11. Zurada JM. Introduction to Artificial Neural Systems. West Publishing Company, New York, 1992
12. Newton RT, Xu Y. Neural Network Control of Space Manipulator, Control Systems, 1993; 13: 6
13. Arena P. Reti Neurali per Modellistica e Predizione in Circuiti e Sistemi', Ph.D. Thesis. Dipartimento Elettrico Elettronico e Sistemistico, Università di Catania, Italy 1993
14. Fortuna L, Muscato G, Nunnari G, Pandolfo A, Plebe A. Application of Neural Control in Agriculture: An Orange Picking Robot. Acta Horticulturae. 1996; 406: 441-450
15. Fortuna L, Muscato G, Nunnari G, Papaleo R. A Neural Networks Approach to Control the Temperature on Rapid Thermal Processing. MELECON 96 IEEE Conference on Industrial Applications in Power Systems, Computer Science and Telecommunications. Bari (Italy), 13-16 Maggio 1996; 649-652
16. Fortuna L, Muscato G, Xibilia MG. Attitude Feedforward Neural Controller in Quaternion Algebra. in Soft Computing with Industrial Applications: Recent Trends in Research and Development. TSI Press, Albuquerque, 1996; 5: 222-228
17. Fortuna L, Gallo A, Muscato G, Xibilia MG. Controllo Neurale in Algebra Quaternionica dell'Assetto di un Corpo Rigido Nello Spazio', AEI Automazione Energia Informazione, UTET periodici. 1997; 84: 5: 72/480-79/487

6. Evolutionary Optimization Algorithms

6.1 Introduction

As pointed out in the previous chapters, both fuzzy logic and neural networks imply *optimization processes*. For fuzzy logic in particular, optimization algorithms are needed that will allow determinations of the number of rules, the number of fuzzy sets and their position in the universe of discourse to be based on optimum criteria instead of on empirical techniques. This process generally involves a large number of variables and thus requires particularly efficient optimization algorithms. Similarly, in the field of neural networks, what can be of considerable use are optimization algorithms capable of finding the *global minimum* of a function with many variables, in order to overcome the intrinsic limitations inherent in learning algorithms based on the gradient technique. Therefore, this chapter will describe *evolutionary algorithms* that seem to respond to the characteristics required by soft computing, both with regard to versatility and to the efficiency and goodness of the results obtained. *Genetic algorithms* have proved to be a valid procedure for *global optimization*, applicable in very many sectors of engineering [10-15]. Ease of implementation and the potentiality inherent in an evolutionist approach make genetic algorithms a powerful optimization tool for *non-convex* functions. The genetic algorithms (GA) represent a new optimization procedure based on Darwin's *natural evolution* principle. Adopting this analogy, inside a *population* in continuous evolution, the individual who best *adapts* to environmental constraints corresponds to the optimal solution of the problem to be solved. For example, in an environment formed exclusively of tall plants, the optimal solution to the problem of adaptation is that of the giraffe, capable of surviving thanks to the conformation of its neck, and certainly not the tortoise whose physical structure is characterized by totally different features. As is well known from the Darwinian theory, the slow lengthening of the giraffe's neck is a process of optimization constrained by environmental cues, and it is precisely this kind of phenomena that inspire genetic algorithms.

Therefore, the analogy is to be found in the following relations:

Population → Environmental cues → Best adapted individuals

Possible solutions to the problem to be optimized → Mathematical constraints imposed → Optimal solutions

Among the various optimization techniques, the point where the GA should be placed can be clearly identified from an analysis of Figure 6.1.

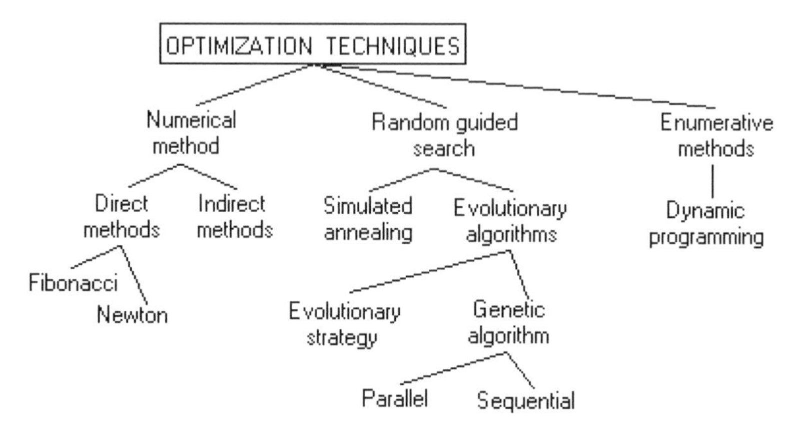

Figure 6.1. Optimization methods

Generally, the optimization algorithms can be subdivided into three large classes: *numerical methods*, *random search*, *enumerative techniques*.

The *numerical methods* are based on a set of necessary and sufficient conditions that are verified by utilizing optimization procedures. In particular, these methods are divided into two subgroups: *direct* and *indirect* methods.

Direct methods consist in the search for solutions to a set of non-linear equations deriving from having the gradient of the function to be optimized set equal to zero.

Indirect methods, like those of Newton or Fibonacci, are *iterative* methods, also based on information supplied by the objective function gradient. Whereas the direct method by solving a system of equations directly determines the optimal point, the indirect method shifts the working point inside the domain of interest, assessing from moment to moment the information deriving from the gradient calculated in the point in question. Although they are the ones most used, these methods do not guarantee conditions of global optima, and furthermore prove to be efficacious only for a set of well-defined problems.

The *enumerative* techniques make the domain of interest discrete and calculate the objective function for all the points considered. They are easy to implement but heavy from a computational standpoint. Dynamic programming is a classical example of enumeration techniques.

The *random search* methods are based on *enumerative* techniques, but, during their execution, also complementary information is used which proves helpful in searching for the optimum. They are of general application and useful in many complex problems. These methods, too, are divided into subgroups: *evolutionary algorithms* and *simulated annealing*, both of which simulate natural phenomena.

Simulated annealing is inspired by a thermodynamic process which simulates the slow cooling of a body, whereas evolutionist algorithms arise from a simulation of the process of natural evolution.

Evolutionary algorithms in their turn can be divided into *genetic algorithms* and *evolutionary strategies*.

While noting the analogy with natural processes, the former act more at a genetic level, whereas the latter place greater emphasis on the behavioral relationship between parents and offspring.

Genetic algorithms display certain characteristics that in some applications make them the only valid alternative to classical optimization methods.

A GA does not process a single point in the domain of interest but instead deals contemporaneously with a *population* of points. In this way it is possible to rise in parallel a multi-modal function thereby avoiding being trapped in local minima.

During their execution, GA use *probabilistic* and not deterministic transition rules, unlike the classical optimization methods to determine the optimum. GA need only to know the objective function.

Thus they do not need to know the n-th derivative or have any further information about the function to be optimized. Apart from the Darwinian approach, this characteristic is perhaps the one which makes these algorithms more efficacious and better alternatives than the traditional optimization techniques.

6.2 The Structure of a Genetic Algorithm

As already stated, unlike the classical methods such as that of the gradient or the simplex one, a GA elaborates a set of points, called populations rather than a single value. All these points belong to the domain of the function to be optimized.
In particular, while the optimization methods like that of the gradient move in a direction determined by the gradient at the point considered, genetic algorithms process *in parallel* a set of points, thereby ensuring that the information characterizing each of them can be transmitted and combined with that of the others.
Generally, the points of the domain of interest are *codified* in structures that recall a *chromosome* conformation.

Also here, an analogy can be made with natural phenomena.

A possible structure, and generally the one most used, is that of a *binary string*.

Let us suppose we have a mono-dimensional case, *i.e.*, we have to optimize a function of one independent variable only, a generic value x^* belonging to the domain of the optimization variable x may be codified as shown in Figure 6.2.

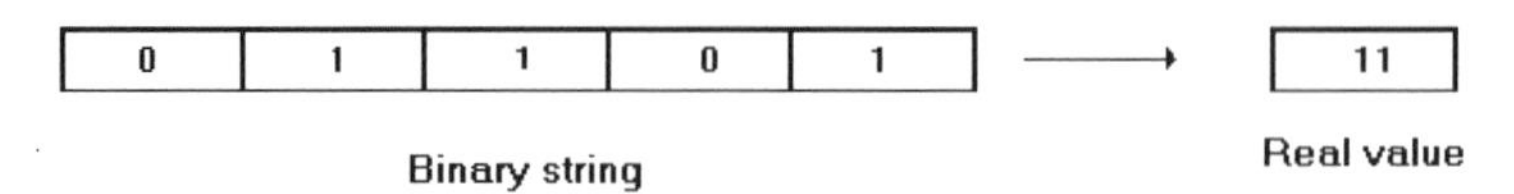

Figure 6.2. Binary codification

In this case, in order to obtain the value of *x*, a simple binary-real conversion has been effected.

By making use of other types of conversion, we can, provided always we start out from binary string structures, both map the points inside a preset domain [X_{min}, X_{max}] and adopt floating point representations by using mantissas and exponents. If we intend to optimize a function having several variables, the so-called *multi-parameter concatenation* is used.

This representation consists in connecting the binary strings representing the individual variables into a single string, automatically identified as being a population element. An example of this is given in Figure 6.3.

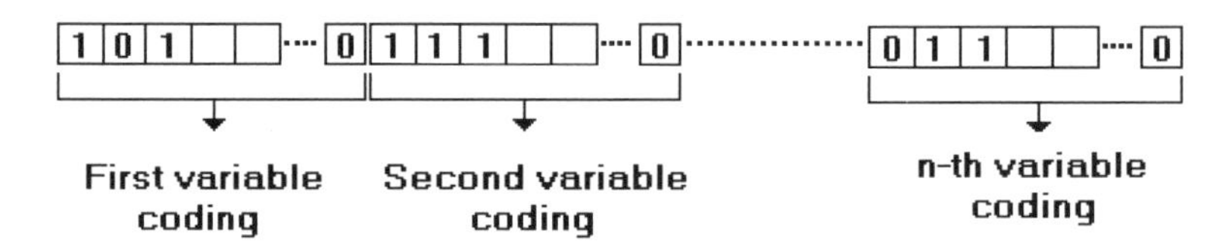

Figure 6.3. Multi-parameter concatenation

During the execution of the program, that is to say during the creation of new *generations*, one or more of the following operators are applied to all population elements:

- *reproduction*: this consists in recopying an element of the current population into a new population which represents the *successive generation*;
- *crossover*: when two strings have been fixed, with the interchange between them of random length sub-strings, two elements of the new generation are created, as is shown in Figure 6.4.
- *mutation*: given a single string, a randomly chosen bit is complemented and the new element thus obtained will be added to the new generation. The action of this operation can be seen in Figure 6.5.

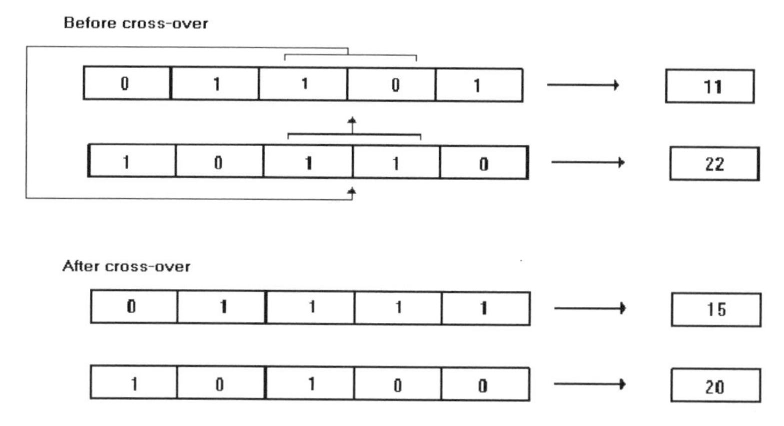

Figure 6.4. Crossover

During the execution of the algorithm, as will be explained more clearly below, the three operators are applied with *probabilities* of p_r, p_c, and p_m, respectively.

All the elements of the population are characterized by an index of goodness indicated as *fitness function*. The fitness of a string is closely connected to the value assumed by the function to be optimized in that particular point.

The higher the fitness value, the more that point, belonging to the domain of the function to be optimized, will be close to the point of global optimum.

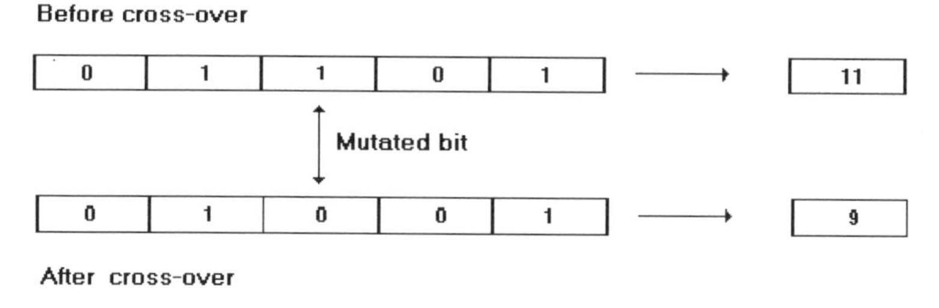

Figure 6.5. Mutation

A distinction should be made between the concept of *objective function* and that of *fitness*. The *objective function* supplies a measurement of the performances with respect to a particular solution, while the *fitness function* transforms this measurement into a new value that will be useful for reproduction. The fitness value is therefore determinant in identifying those elements of the population that are the most promising ones, *i.e.*, those which in an evolutionist sense are the *best adapted* and that thus have the greatest probability, when they reproduce, to hand down their genetic heritage to successive generations. In fact, by choosing the elements of the population in a probabilistic way according to the fitness value, the best elements of the population are identified and the three previously described operators are applied to them.

On the whole, a GA can therefore be represented schematically as shown in Figure 6.6.

With the turnover of generations, only those elements with a high fitness value will be able to reproduce, by means of selection, and hand down their genes (bit) by crossover. These elements represent the variables whose objective function value is closest to the optimum. At the conclusion of this process, the element with the *highest fitness*, that is to say the most fit one, corresponds, within the limits of the pre-defined stop criterion, the optimal solution to the problem.

To terminate the algorithm, several *stop criteria* can be used. The simplest consists in fixing the *maximum number of generations* and electing, at the end of the algorithm, the element with the highest fitness as being the optimal solution.

Another technique consists in monitoring the *mean* and *maximum fitness values*, and when the difference between these two values has reached a preset value, the algorithm is automatically blocked. In this case, all the elements of the population are by then close to the optimal value so that the mean and maximum values coincide.

Let us analyze in particular the effect of the three previously mentioned operators. The *reproduction* operator is used for increasing, in the populations, the elements with the highest fitness, which is to say the most promising ones.

The *crossover* operator is used for combining the genetic information belonging to different parents, thereby allowing the transfer of the best characteristics to the strings of a new generation.

Finally, the *mutation* operator adds new elements to the populations.

In particular, the mutation of a bit, even though it may cause considerable variation in the fitness value of a string, allows us to avoid *local minima*, while if a small variation is involved, it allows a numerical refining of the function that has to be optimized around the point reached.

We have previously mentioned p_r, the reproduction probability, p_c, the crossover probability, and p_m, the mutation probability. Let us now see what this means.

It should first of all be pointed out that these parameters are preset quantities, and that generally they remain constant during the whole optimization process. The literature reports approximate values to associate with the three probabilities, but it has been found that when varying the application it is convenient, in order to improve the algorithm's performance, to vary the values of these parameters as well.

To *tune* the parameters, one must thus make use of the user's experience.

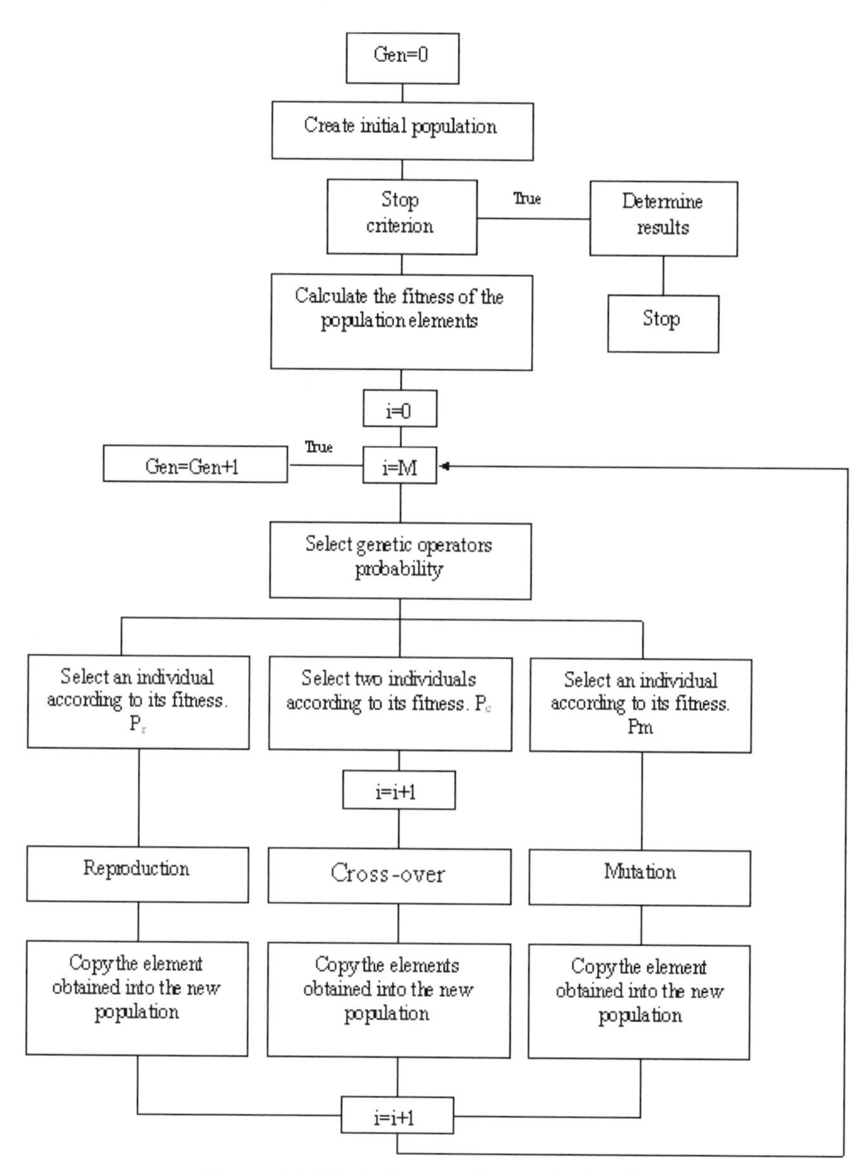

Figure 6.6. Block diagram of a genetic algorithm

Whenever a well-adapted element is chosen according to its fitness value, it must be decided which fitness value to apply to it. Therefore, we generate a random number p^*, ranging between zero and one; if $p_r \leq p^*$, then the element considered will be chosen such that it makes part of the new population.

If instead $p_r > p^*$, the element will be discarded and will not form part of the new population being created.

The procedure for the crossover and mutation operators is completely analogous, the only difference being that before applying the crossover two strings must be chosen, and not just one.

The application of the described operators, leading exclusively to the generation of random numbers, the copying, and the exchange of binary strings all make GA extremely simple to implement.

Another expedient adopted for improving the performance of a GA is the so-called *élitism*. As generations pass, this technique consists in maintaining, by recopying it, the element with the highest fitness.

Given the effect of crossover, reproduction, and mutation, there is in fact no guarantee, in a simple genetic algorithm, that the best-adapted individual will survive.

When implementing the algorithm, particular attention should be addressed to the fitness of each element of the population. Generally, fitness is directly linked to the value assumed by the objective function. However, when determining fitness, we can use routines that take constraints into account by applying corrective terms to the fitness value.

6.3 A Numerical Example

Let us consider a population made up four strings, each of which with five bits, and let us optimize, or in this case maximize, the objective function:

$$f(x) = 2 * x$$

Given a binary string, the previous function, which we will suppose for convenience to be defined in $0 \le x \le 31$, associates the binary string with the corresponding value of x, *i.e.*: $f(00000)=0$, $f(00001)=2$, $f(00010)=4$, *etc*. The initial population, as said previously, is created randomly, generating a set of points belonging to the considered domain.

The first five columns of Table 6.1 report, respectively:

- the number of strings;
- the binary coding;
- the fitness, calculated in this case as: $f = f(x_i) = 2 * x_i$;
- the percent total fitness associated with each string:

$$\frac{f_i}{\sum f_i}$$

The estimated number of copies of an element in the successive population is given by:

$$\frac{f_i}{f_{mean}}$$

The value reported in the fourth column indicates the probability of the individual element of the population being chosen for application of the three operators.

This value is clearly linked to the fitness value. The point represented by 11000 thus has a 38.1/100 probability of being chosen to form part of the successive population, whereas the point codified by 00101 has a 7.9/100 probability, and so forth.

The selection process can be activated by a turn of the *weighted roulette wheel*, that is to say a roulette wheel with sectors of width proportional to the fitness, like that reported in Figure 6.7.

By employing that wheel, the elements have been chosen which will create the successive generation; in particular, the last column of Table 6.1 indicates how many times that element has been chosen to make part of the new population after being processed by the crossover and mutation operators. From these values, which during the execution of the algorithm are monitored merely for statistical purposes, it can be noted that String 1, which has the highest fitness, is chosen twice, whereas Strings 3 and 4 are selected once only, and String 2, having a low fitness value, is discarded. A new population is thus created by identifying only the most promising elements and applying to them the reproduction, crossover, and mutation operators.

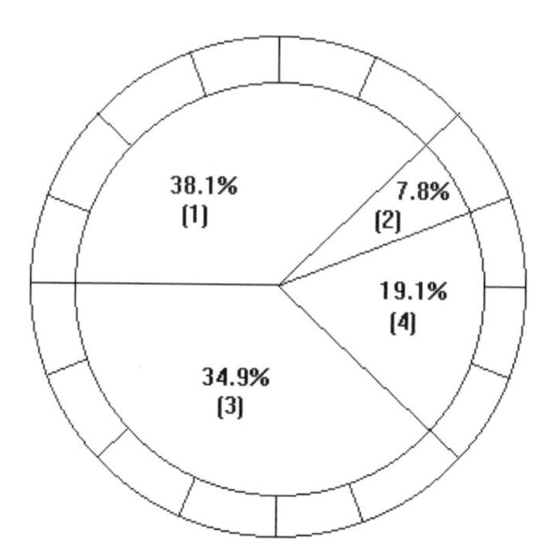

Figure 6.7. A roulette wheel with sectors proportional to the fitness

Table 6.1. Initial population and characteristic parameters

i	String x_i	Fitness $f(x_i)=2*x$	$\frac{f_i}{\sum f_i}$	Estimated number of copies $\frac{f_i}{f_{mean}}$	Real number of copies
1	11000	48	0.381	1.524	2
2	00101	10	0.079	0.317	0
3	10110	44	0.349	1.397	1
4	01100	24	0.191	0.726	1
	Total	126	1.0	4.0	4
	Average	31.5	0.250	1.0	1
	Max	48	0.381	1.524	2

In Table 6.2 it can be seen that in only one generation the average fitness has increased by 30% and the maximum value by 25%.

Table 6.2. Population after selection and crossover

After reproduction	Associate string	Cross-over point	After cross-over	Fitness of $f(x_i) = 2\ x_i$
11-000	x3	2	11110	60
1-1000	x4	1	11100	56
10-110	x1	2	10000	32
0-1100	x2	1	01000	16
Somma				164
Media				41
Max				60

This simple process is reiterated until, as said already, a fixed stop criterion is verified. Once the algorithm has been blocked, the individual 11111, obtained after five generations, represents the optimal point for the objective function.

A further example is reported below, showing the capacity of GA to process in parallel all the points of the population when calculating the maxima (minima) of the functions to be optimized.

The function in question, the well-known $y=sin^2(x)/x$, can clearly be maximized using standard optimization methods, and it has been chosen exclusively for graphic reasons in so far that it allows us to see how the possible solutions evolve by being shifted.

Figure 6.8 shows the function to be maximized, while Figures 6.9 through 6.12 show the solutions with the varying of generations.

In particular, the asterisks plotted on the function represent the phenotypes, that is to say the values assumed by the parameters to be optimized with varying of generations and under the effect of the genetic operators.

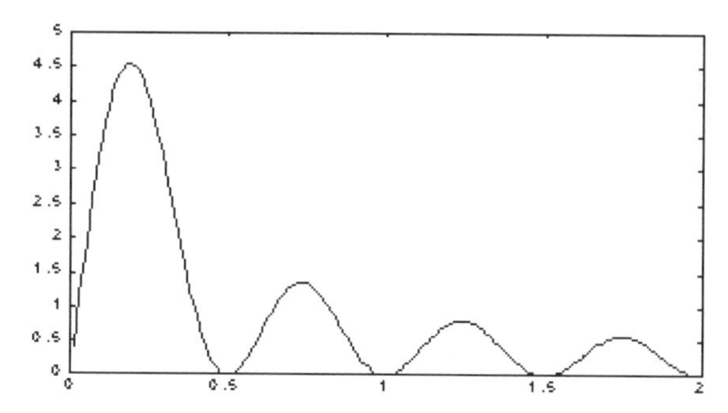

Figure 6.8. Graph of the function

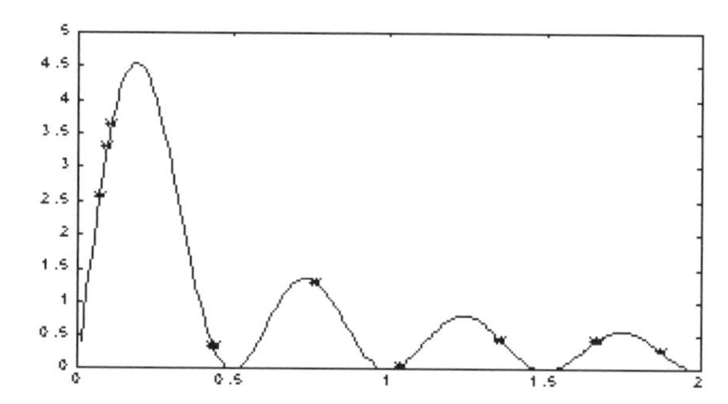

Figure 6.9. Population phenotype distribution ($f(x)$) in the initial population

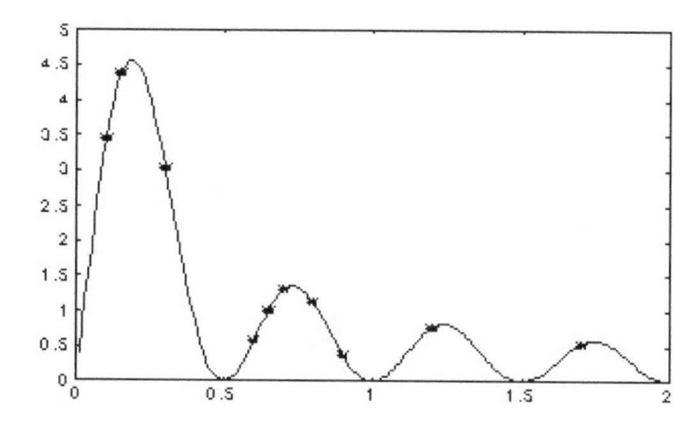

Figure 6.10. Population phenotype distribution ($f(x)$) at the fifth generation

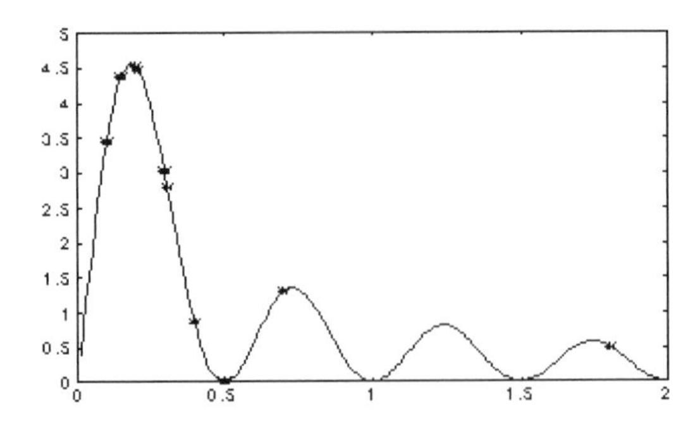

Figure 6.11. Population phenotype distribution ($f(x)$) at the tenth generation

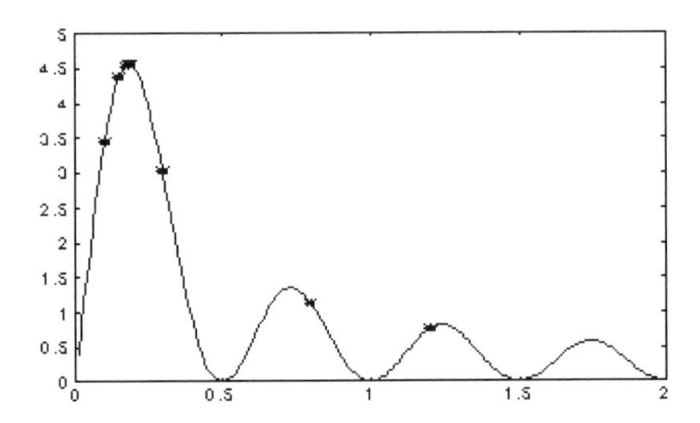

Figure 6.12. Population phenotype distribution ($f(x)$) at the fifteenth generation

6.4 The Schema Theorem

The suggestive analogy with Darwin's theory and ease of implementation could make GA seem more of an *optimization software game* than a valid alternative to the traditional optimization methods. A rigorous theoretic discussion of GA was, however, offered first by Holland and later by Goldberg [1], [2].
With a view to formalizing what has been said above, some definitions should be introduced:

Definition 6.1. *Adopting a binary representation in codifying the variables to be optimized, a scheme is defined as a "similarity in form" between strings of the type:*

$$H = (b_1, b_2, b_3, \ldots, b_l) \text{ with } b_l \in (0,1,*)$$

*where the symbol * is a symbol which can assume both the value 1 and the value 0.*

Given, for example, the binary strings

A=1 0 1 0 0 1 0

B=1 1 0 1 1 1 1

it can be observed that both have a 1 in the first and in the second position. The similarity in the form can therefore be formalized by adopting the following structure:

H=1 * * * * 1

Thus this scheme characterizes not just two strings A and B, but 2^5 different strings having the value 1 fixed in the first and sixth positions. It is said that these two strings are part of the 2^5 strings that belong to, or give an instance of, that scheme.

Definition 6.2. *The number of fixed positions in the considered scheme is defined as the order of a scheme H;*

for example forScheme 1.3, it is true that:

$$\sigma(H)=\sigma(1****1*)=2$$

Definition 6.3. *The length of a scheme H is given by the distance between the two extreme positions fixed in (1.3)*

the following is valid:

$$\delta(H) = \delta(1****1*) = 6 - 1 = 5$$

Once these characteristic sizes have been introduced, the basic GA theorem, the so-called Schemes Theorem, can be introduced.

Theorem 6.1. *Let us suppose that the reproduction occurs in a manner proportional to the fitness, under the effect of crossover and mutation, then the number of copies m of a scheme in the instant t+1 is determined by the following expression:*

$$m(H,t+1) \geq m(H,t)\frac{f(H)}{\bar{f}}\left[1 - p_c\frac{\delta(H)}{l-1} - p_m\sigma(H)\right]$$

where f(H) represents the fitness of the scheme H (i.e., the average fitness of all the population strings indicated in scheme H), $\bar{f}$ is the mean fitness value of the elements forming the whole population, while p_c and p_m represent the previously defined probability of crossover and mutation, respectively.

The value of f(H) is obtained by means of:

$$f(H) = \frac{\sum_{i \in H} f(i)}{m(H,t)}$$

where f(i) is the fitness of the i-th element belonging to scheme H.

A direct consequence of the schemes theorem is that schemes of no great length and of low order have an exponential probability of being represented in following generations. At this point we can introduce a basic concept regarding GA From what has been stated so far, it can be easily understood how, in manipulating a string population, the search is shifted in reality from the space of the strings to that of the *similar strings*, which is to say the *schemes*. If we adopt a binary coding, a population of *n* different strings includes from *2* to *2*n* schemes and, bearing in mind that such a high number of schemes is processed in parallel, while computationally one operates with a far smaller number of elements, we can appreciate the potentiality of this new approach to optimization problems. This concept is generally defined as *intrinsic parallelism*. In the last two years, together with the applications, the theoretical aspect of GA has been studied in greater depth, characterizing the salient features regarding the convergence of the algorithm. Although at first this aspect aroused little interest, in so far that a *good solution* in terms of time and acceptable costs was preferred, today an in-depth theoretical discussion has led to the academic recognition of GA A thorough treatment of the problem of GA convergence is something outside the aims of this chapter, but some bibliographical references will be given. The approaches followed for formulating the *convergence theory* have been many but in general the one currently most adopted is connected with probabilistic models. In [3-7], by exploiting the *Markov chains theory*, the convergence of a canonical form of genetic algorithm has been proved. In particular, it was proved that a genetic algorithm implemented adopting the élitism technique converges towards the global optimum. Other theorems have been developed considering particular types of algorithms or specific classes of objective functions; at present, however, although considerable efforts have been addressed to studying convergence, ever greater attention is being paid to characterizing the computational load, in the temporal sense, of genetic algorithms.

6.5 How to Improve the Performance of a Genetic Algorithm

The prototype of genetic algorithm described in the previous sections is primitive in form. That is to say it corresponds with the standard algorithm reported in literature.

Naturally, very many variations have been proposed which have led to *improved performances* of the basic algorithm. Ideally, attempts have been made to create a general purpose genetic algorithm capable of solving a vast variety of problems; but in reality the innovations proposed prove useful for some applications and less so for others.

The first step in measuring the performances of an algorithm consists in creating a *test criterion*. In De Jong, five functions were proposed, having particular characteristics, which should be taken into account every time a new algorithm is developed [8].

The functions *f1, ..., f5*, reported below, have the characteristic of being discontinuous, multi-modal and non-convex; thus they prove to be an excellent test for any optimization algorithm:

$$f_1(x_i) = \sum_{1}^{3} x_i^2$$

$$f_2(x_i) = 100(x_1^2 - x_2)^2 + (1 - x_1)^2$$

$$f_3(x_i) = \sum_{1}^{5} \text{int}(x_i)$$

$$f_4(x_i) = \sum_{1}^{3} 0ix_i^4 + Gauss(0.1)$$

$$f_5(x_i) = 0.002 + \sum_{j=1}^{5} 5 \frac{1}{j + \sum_{i=1}^{2} (x_i - a_{ij})^6}$$

In order to quantify the goodness of an algorithm, two performance indices have been introduced, called, respectively:

$$\textit{on-line performance: } P_{on} = \frac{1}{T} \sum_{1}^{T} f(t)$$

$$\textit{off-line performance: } P_{on}^* = \frac{1}{T} \sum_{1}^{T} f^*(t)$$

These represent, respectively, the average value of fitness f at the T-th generation, and the average value of fitness f^* of the best string of the population, again at the T-th generation.

In De Jong, apart from the test functions, variations of the basic algorithm were proposed. The techniques employed are the following:

- élitism;
- a model with a foreseen number of copies;
- crossovers at several points.

These are characterized, in particular, as follows:

The technique of élitism, as previously mentioned, consists in preserving the survival of that element with the highest fitness at the passing of generations. Let $a^*(t)$ be the element with the highest fitness in the generation $A(t)$. If, after creating $A(t+1)$, the element $a^*(t)$ is not part, then it is added as the $(n+1)^{th}$ element of the current population. This approach has proved to be determinant in optimizing discontinuous functions, whereas it has been found less indicated in cases of multi-modal functions.

In order to reduce the intrinsic stochastic error in the roulette wheel selection method, the number of foreseen copies method has been employed, which is based on the following considerations. Often, when adopting the roulette wheel method, it happens that, due to stochastic errors, for one element of the population the number of foreseen copies and the effective one do not coincide. It is precisely to solve this problem that the model of foreseen copies technique has been introduced.

Crossover at one point only, *i.e.*, the one previously described, is not able to recombine schemes having particular characteristics. As can be noted from the following example, applying a one point only crossover to the two elements *a1* and *a2*, the scheme reported in boldface cannot be combined.

a1 = **1**1**0**11001011011**1**

a2 = 0001**0**110**1**11100

If, instead, a two-points crossover is employed, the scheme can be recombined by effecting the following double crossover:

a1= **1**1**0**1|100101|101**1**

a2=0001|**0**110**1**1|1100

obtaining

b1= **1**1**0**1**0**110**1**1101**1**

b2=00011001011100

where in *b1* the desired scheme has been reconstructed.

This method proves unproductive if a large number of crossover points is introduced. In that case, in fact, there is the risk of breaking the scheme into too many points, with only slight chance of reconstructing it in the successive generations.

Another characteristic genetic algorithm parameter is the number of elements forming the population. It is wrong to think that choosing a large number of elements will help improve the algorithm's characteristics. In fact, if on the one hand a greater amount of information regarding the domain of the function to be optimized is available, on the other the algorithm's computational load is considerably enlarged. Again on the other hand, too small a population will prove unsatisfactory because in a limited number of generations all the strings of the population would tend to assume the same structure of zeros and ones, and would thus represent one point only.

It is reported in the literature that if the chromosome length of a generic element of the population is equal to m it is advisable to fix the maximum number of elements forming the population at approximately $2*m$.

Another factor to be considered for improving the performance of a genetic algorithm is the following. Often we already have some information regarding the domain of the function that is to be optimized. Thus we are aware of the fact that some zones of the domain are more promising than others in searching for the optimum. Hence the so-called seeding technique arises. This method consists in putting a *seed* or several seeds into the initial population. Thereby additional information is supplied to the algorithm, indicating in which zones of the domain the research should be focused.

During a genetic algorithm optimization process, one problem that must necessarily be avoided is the so-called *premature convergence*. We will now see in what it consists. If during the first generations, a string with fitness considerably higher than that of the others is present among the elements of a population, and if the evolutionary pressure is very strong, that is to say a high probability value for reproduction and crossover, then in a few generations this excellently equipped individual will succeed in dominating, leading to a premature convergence of the algorithm towards a point which will unlikely be the global optimum.

Instead, the dual problem arises during the final phase of the optimization process. In fact, during this phase the population is formed of points that are very close to each other and near the global optimum. In this condition, given that the fitness of the strings is practically the same for all, it is hard for the algorithm to identify merely by the roulette wheel technique the points closest to the global optimum.

A possible method for solving this problem is that of redefining the fitness function by means of a linear relation of the type:

$$f' = af + b$$

The coefficients a and b can be chosen in various ways, but generally it is desirable that $f'_{avg} = f_{avg}$ and $f'_{min} = f_{min}$. To check the number of copies that the element with highest fitness will produce in the successive generation, the following relation is needed:

$$f'_{\max} = C_{\text{mult}} * f_{\text{avg}}.$$

For those populations made up of at most 100 elements, the multiplication factor C_{mult} is generally set equal to 2. Using this procedure, it might happen that elements with a low fitness f, after applying the previous relations, assume negative f' values.

To avoid that, conditions must be imposed on the coefficients a and b such that the condition $f'_{min} = 0$ is respected. What has been stated can be easily implemented by employing the following statistical parameters of the genetic algorithm: average, minimum, and maximum values of fitness.

Another problem that can be readily overcome is that connected with a simple binary number $\rightarrow$ whole number coding. In fact, the two numbers 31 and 32, although successive ones, have a substantially different binary representation:

$$31 = 011111$$

$$32 = 100000$$

To solve this problem, several techniques are available. The simplest consists in using the *Gray code* rather than the binary one.

Instead, another approach is to use real numbers directly and apply to them the suitably adapted crossover and mutation operators. This kind of representation has been opposed by several authors because in part it makes the chromosome representation unnatural and therefore does not offer theoretical support for the schemes theorem.

To enhance GA efficiency, certain variants for the classical algorithms can be created which exploit much more the naturalistic aspect of genetic algorithms. The concept of implicit parallelism has already been defined, and on the basis of this presupposition, very many parallel processor structures have been proposed on which to evolve a genetic algorithm.

The one presently come to the fore is the island one. A grid of processors is created and on each of these a GA is placed. At pre-established intervals of time, the so-called *migration* is carried out, consisting in an exchange between the various islands of the best elements in the population. Adopting this technique, which of course implies as an optimal solution the best element of all those in the various islands, the algorithm has been considerably speeded up.

Genetic algorithms can deal with *constrained optimization problems*. When the constraints are expressed in the form of inequality, they are generally included in the objective function by means of a penalty function which expresses the degree of violation of the constraint; for example, the problem:

$$\min[f(x)] \ with \ g_i(x) \geq 0 \qquad i = 1,2,...n$$

is expressed with the objective function:

$$\min[f(x)] + r^* \sum_{i=1}^{n} \phi[g_i(x)]$$

where Φ is the penalty function and r is a weight depending on the constraint.

Constraints can also be expressed by means of a suitable gene coding that will allow the points not admissible in the domain of interest to be excluded *a priori*. With procedures such as this, multi-objective optimization problems [16] or those with variable objective function can be solved.

As proved by the ever-growing number of scientific papers and international congresses, genetic algorithms have been definitively accepted in various scientific disciplines as an efficient optimization method, complementary to, and not in competition with, traditional optimization methods. From a starting phase, still today on-going, characterized essentially by the formal development of algorithms, we have moved on to a massive application of these procedures and their integration, within the framework of soft computing, with neural networks and fuzzy logic. The powerful low-cost hardware of today's PCs makes GA readily usable even in problems of medium-sized optimization.

6.6 References

1. Holland J.H., Adaptation in Natural and Artificial System. University of Michigan Press, 1975
2. Goldberg D.E., Genetic Algorithms in Search, Optimization and Machine Learning. Addison Wesley, 1989
3. Nix A., Vose M., Modelling Genetic Algorithm with Markov Chain. Annals of Mathematics and Artificial Intelligence. 1992; 5: 79-88
4. Eiben A., Aartas E., van Hee K., Global Convergence of Genetic Algorithms: a Markov Chain Analysis. Parallel Problem Solving from Nature. Springer Verlag, Berlin, 1990
5. Rudolf G, Convergence Analysis of Canonical Genetic Algorithms. IEEE Transaction on Neural Networks. January1994: 5(1): 96-101
6. Qi X., Palmieri F., Theoretical Analysis of Evolutionary Algorithms with an Infinite Population Size in Continuous Space. Parts I and II. IEEE Transaction on Neural Networks. January 1994; 5(1): 102-129
7. Davis T., A Simulated Annealing like Convergence Theory for Simple Genetic Algorithms. Proceedings of Fourth International Conference on Genetic Algorithms. 1991; 78-181, Morgan Kaufmann, San Matteo, CA
8. De Jong K.A., An Analysis of the Behaviour of the Class of Genetic Adaptive System, Ph. D. thesis Michigan, 1975
9. D.E. Goldberg, A meditation on the application of Genetic Algorithms, IlliGAL Report No. 98003. Department of General Engineering, Universisty of Illinois at Urbana-Champaign
10. Caponetto R., Fortuna L., Graziani S., Xibilia M.G., Genetic algorithms and applications in system engineering: A Survey, Trans. Inst. Measurement and Control, London, 1933; 15: (3): 143-156
11. Caponetto R., Fortuna L., Xibilia M.G, MLP optimal topology via genetic algorithms, Proc. of Int. Conf. on Neural Networks and Genetic Algorithms, Innsbruck, Austria, 1993; 670-674
12. Caponetto R., Fortuna L., Manganaro G., Xibilia M.G., Synchronization-based non linear chaotic circuit identification, SPIE's Int. Symp. on Information, Communications and Comp. Technology, Applications and Systems, Chaotic Circuits for Communications, Philadelphia, USA, October 1995; 48-56
13. R. Caponetto, L. Fortuna, G. Muscato, M.G. Xibilia, Controller Order Reduction by using Genetic Algorithms, Journal of System Engineering. Springer Verlag, London, 1996; 6: 113-118
14. Caponetto R., Fortuna, Manganaro G., Xibilia M.G., Chaotic System Identification via Genetic Algorithm, Proc. of First IEE/IEEE Int. Conf. on Genetic Algorithms in Engeneering Systems: Innovations and Applications (GALESIA '95), Sheffield, UK, September 1995; 170-174
15. Caponetto R., Diamante O., Fortuna L., Muscato G., Xibilia M.G., A Genetic Algorithm Approach to Solve Non-Convex LMI Problems. Int. Symposium on the Mathematical Theory of Networks and Systems. MTNS '98, Padova, Italia, Luglio, 1998

16. Genetic Algorithms in Engineering Systems, edited by A.M.S. Zalzala and P.J. Fleming, Published by IEEE. London, UK, 1997

7. Cellular Neural Networks

7.1 Introduction

Cellular neural networks (CNN), first formulated by L.O. Chua, made their appearance in 1988 [1].

They constitute a particular type of artificial neural network which gave rise to certain innovative aspects such as *parallel time continuous* (analog circuits) *asynchronous processing*, deriving from the local interactions of the network elements (cells).

Subsequent to its first formulation, Chua's original CNN underwent various generalizations and the field of applications of such systems has been enormously widened, so that they are now used in totally different environments to the original ones.

They are now applied in robotics, in solving partial differential equations, associative memories implementation, data processing for biomedical application, simulating new physical and biological systems, and, what is more, these systems are now enjoying an excellent field of application in image processing [2-3]. In addition, it will be seen in the following chapter how suitable cellular neural networks with a reduced number of cells can be successfully employed in complex dynamics modeling [19-21].

Today, cellular neural networks are an important tool for soft computing.

At the same time, soft computing techniques are determinant in designing these complex systems, and in particular in defining templates, precisely on account of the lack of analytical tools for obtaining a generalized procedure for characterizing them.

7.2 The Architecture of Cellular Neural Networks: The Electrical Model

The basic unit of a CNN is called a *cell* and it is implemented by means of an elementary *electric circuit* made up of *linear elements* such as capacitor, resistances, independent and driven generators, and *non-linear* ones that are also implemented with non-linear piloted generators.

Every individual cell is able to *interact* directly with the cells adjacent to it and indirectly with all the others, by exploiting the effect of stimulus propagation [1-2].

A CNN can be considered a matrix of n-dimensional cells: in our case, we will only be concerned with a two-dimensional CNN, *i.e.*, $M \times N$ matrices, bearing in mind that what will be said can be easily extended to cases of more than two dimensions.

An example of a two-dimensional 3×3 CNN in which the connections between the cells are indicated is shown in Figure 7.1.

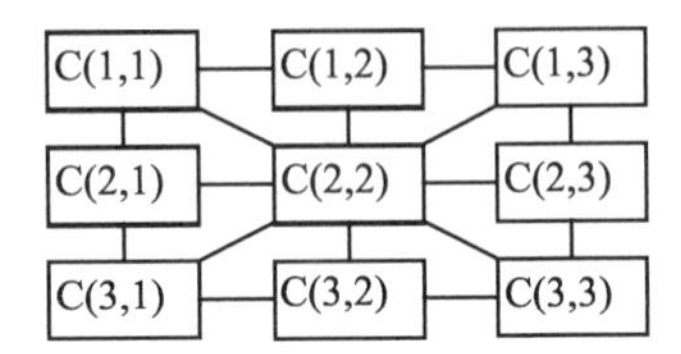

Figure 7.1. Scheme of a 3×3 CNN

The CNN matrix cell, $C(i,j)$, belonging to the i-th row and the j-th column, is defined neighbourhood $Nr(i,j)$ with radius r of the cell, as follows:

$$N_r(i,j) = \{C(k,l) \mid \max(|k-i|,|l-j|) \leq r, \quad 1 \leq k \leq M, \quad 1 \leq l \leq N\} \quad (7.1)$$

where $r > 0$.

Figure 7.2 shows two examples of a cell neighbourhood $C(i,j)$, indicated in black with radii of r=1 and r=2, respectively.

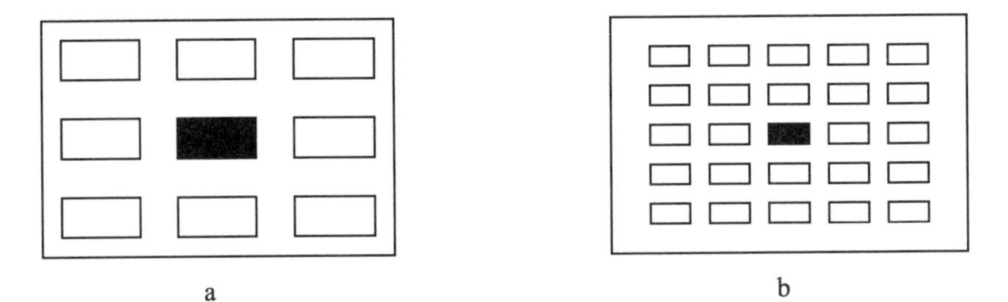

Figure 7.2. Neighbourhood of: a. radius 1 and b. radius 2 of a cell

It should be noted that a 3×3 neighbourhood corresponds with a r=1 radius, and a 5×5 one with a r=2 radius, and so on.

Moreover, again indicating the generic cell by $C(i,j)$, for a CNN the following symmetry is valid:

$$\text{If } C(i,j) \in Nr(k,l), \text{ then } C(k,l) \in Nr(i,j), \quad (7.2)$$

$$\forall C(i,j), \forall C(k,l),$$

Figure 7.3 reports the electrical equivalent circuit to a generic CNN cell. The suffixes u, x, and j indicate, respectively, the input, state, and output of the circuit.

The only non-linear element whose characteristic can be presumed to be piece-wise linear is the voltage driven generator *Iyx*, whose characteristic is reported in Figure 7.4. Hence, the following relation holds:

$$f(V)=\frac{1}{2}[|V+1|-|V-1|] \tag{7.3}$$

Given this, the following equations (normalized with respect to a unitary output voltage $|Vyij| \le 1 volt$) are valid for the current generators:

$$Iyx(i,j)=(1/Ry)*f(Vxij)=(1/2Ry)*(|Vxij+1|-|Vxij-1|) \tag{7.4}$$

$$Ixy(i,j;k,l)=A(i,j;k,l)*Vykl=\frac{1}{Ry}*A(i,j;k,l)*Iyx(k,l) \tag{7.5}$$

$$Ixu(i,j;k,l)=B(i,j;k,l)*Vukl \tag{7.6}$$

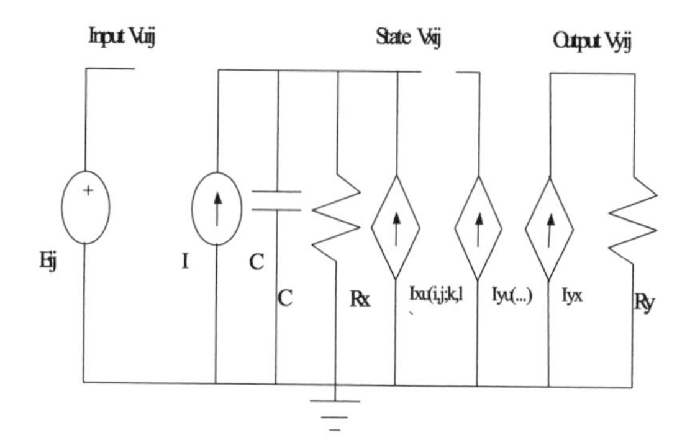

Figure 7.3. Electrical scheme of a CNN cell

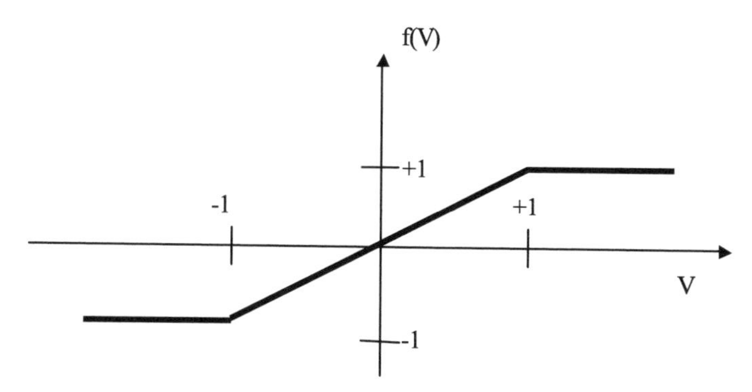

Figure 7.4. Characteristics of the non-linear current generator

It can be noted that the interaction between the cell *C(i,j)* and the cell *C(k,l)* occurs by means of voltage driven generators.

The matrices A and B are called convolution masks or templates: together with the current I, they affect the behavior of the entire CNN. A necessary condition for convergence toward a stable state of the CNN is that A be a symmetrical matrix, *i.e.*:

$$\forall k, l \in Nr, \text{ is: } A(k, l) = A(l, k) \tag{7.7}$$

Moreover, if $A(0,0) > 1$, every stable state output will fall between the values of 1 and -1.

The template A is called a *feedback template*, since it is linked to the circuit outputs and performs an effective state feedback action.

With regard to template B, it can be interpreted as a *control operator* and is therefore called a *control template*, since it is moreover connected to the circuit inputs [4-5]. The block diagram of a standard CNN is given in Figure 7.5:

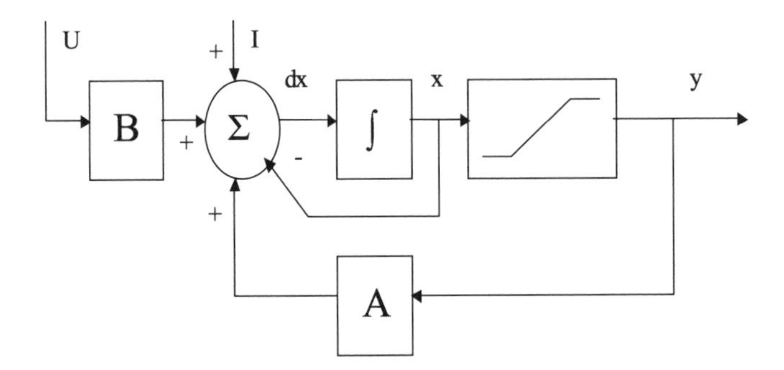

Figure 7.5. Block diagram of a CNN

By applying the laws of Kirchhoff regarding current and voltage to the equivalent circuit in Figure 7.3, the following state equations are obtained for the cell C(*i,j*) of a CNN:

State Equations:

$$C\frac{d}{dt}Vxij(t) = -\frac{1}{Rx}Vxij(t) + \sum_{C(k,l)\in Nr(i,j)} A(i,j;k,l) * Vykl + \sum_{C(k,l)\in Nr(i,j)} B(i,j;k,l) * Vukl + I \tag{7.8}$$

where $1 \leq i \leq M$ and $1 \leq j \leq N$ and where M and N indicate respectively, the number of rows and the number of columns of the CNN.

Output Equations:

$$Vyij(t) = \frac{1}{2}(|Vxij(t)+1| - |Vxij(t)-1|) \tag{7.9}$$

where $1 \leq i \leq M$ and $1 \leq j \leq N$

Input Equations:

$$Vuij = Eij\text{h} \tag{7.10}$$

where $1 \leq i \leq M$ and $1 \leq j \leq N$

Constraints:

$$|Vxij(0)| \leq 1 \tag{7.11}$$

$$|Vuij| \leq 1 \tag{7.12}$$

where $1 \leq i \leq M$ and $1 \leq j \leq N$

Parameter Constraints:

$$A(i, j; k, l) = A(k, l; I, j) \tag{7.13}$$

$$C > 0 \tag{7.14}$$

$$Rx > 0 \tag{7.15}$$

where $1 \leq i \leq M$ and $1 \leq j \leq N$

The following features should be mentioned:

- denoting with $C(i,j)$, a generic CNN inner cell; the cells bordering on it are called the neighbouring, or boundary, cells, and their number, once the neighbourhood radius of $C(i,j,)$ is indicated with r, will be equal to $(2r + 1)^2$;
- every cell of a CNN has at the most three nodes (as is seen in the circuit of Figure 7.3); in fact, selecting, for example, $Eij = 0$, if $B(i,j;k,l) = 0$, for all the cells the nodes of each one will be reduced to two (state and output);
- the elements of templates A and B have the function of weights for the state equations and act as effective convolution operators carrying out, respectively, an output feedback and an input control;
- the resistance values Rx and Ry represent the loss of Joules in the circuit, with reference to the circuit of Figure 7.3, and have values normally falling in the range between one thousand and one million ohms;
- the time constant of the circuit, *i.e.*, $\tau = RC$, has values ranging between one hundred picaseconds and ten microseconds, a feature which makes CNNs suitable for uses such as image processing, since it allows real time processing (continuous time, since here the circuits are analog networks.

7.3 Stability of a Cellular Neural Network

One of the basic properties of CNNs is their characteristic, in feedback template symmetry hypotheses, to converge on a *stable state*, on the basis of certain initial conditions which we will analyze in detail below [1], [6].

When a CNN is applied, for example, to the processing of an image, starting from the original input image it progressively maps all the pixels of the given image and, depending on the coefficients of A and B and of the current I of relation (7.8), it supplies a system of n differential equations that make up the state equations of the pixel cell on which in that moment the CNN is centered (n being the number of cells contained). Whatever may be the initial state of the transitory regime of an individual cell's evolution, the starting image should converge on an output one representing a stable state. We will distinguish below between the stable state of the individual cell and that of the whole CNN, the latter being the main focus of our interest.

Now we will see which and how many are the necessary conditions for the CNN to reach stable state convergence. The study of CNN stability will, for the sake of simplicity, be limited to the case of binary images (values of output tension *Vyij*) with pixel intensity levels of -1 and +1; however, such a study could easily be extended to the case of output images with several gray levels.

Since the cells of the network are characterized by non-linear differential equations, we will employ the Lyapunov convergence criterion for non-linear dynamic circuits. In his articles, L.O. Chua defined the Lyapunov function for a CNN with the following expression:

$$E(t) = -\frac{1}{2}\sum_{(i,j)}\sum_{(k,l)} A(i,j;k,l) * Vyij(t) * Vyki(t) + \frac{1}{2Rx}\sum_{(i,j)} Vyij^2(t) \tag{7.16}$$
$$-\sum_{(i,j)}\sum_{(k,l)} B(i,j;k,l) * Vyij(t) * Vukl(t) - \sum_{(i,j)} I * Vyij(t)$$

Expression 7.16 does not have any precise physical significance: we will merely consider it a form of energy in the generalized sense. It should be noted how in (7.16) the *Vxij* state of the cell does not appear; despite that, however, information on the stable state CNN can still be obtained from it. It is shown that the above-defined Lyapunov function always converges to a local minimum in which the CNN produces the output desired. Now we will examine the properties of expression:

Theorem 7.1. *The Lyapunov function E(t) is upper bounded, i.e.:*

$$\max_t |E(t)| \le E_{\max} \tag{7.17}$$

where $E_{\max}$ has the following expression:

$$E_{\max} = \frac{1}{2}\sum_{(i,j)}\sum_{(k,l)} |A(i,j;k,l)| + \sum_{(i,j)}\sum_{(k,l)} |B(i,j;k,l)| + M * N * (\frac{1}{2Rx} + |I|) \tag{7.18}$$

for a CNN of M rows and N columns.

For integrated circuits, this E_{max} is typically a value around 20 V.

Theorem 7.2. *The Lyapunov function E(t) is a decreasing celly monoton function, as shown here:*

$$\frac{d}{dt}E(t) = -\sum_{|Vxij|<1} C * \left[\frac{d}{dt}Vxij(t)\right]^2 \leq 0 \tag{7.19}$$

or

$$\frac{d}{dt}E(t) = -\sum_{|Vxij|<1} C * \left[\frac{d}{dt}Vyij(t)\right]^2 \leq 0 \tag{7.20}$$

or

$$\frac{d}{dt}E(t) = -\sum_{(i,j)} C * \left[\frac{d}{dt}Vyij(t)\right]^2 \leq 0 \tag{7.21}$$

$$\frac{d}{dVxij}Vyij \geq 0$$

The following theorem is derived from Theorems 7.1 and 7.2:

Theorem 7.3. *For the Lyapunov function E(t), for every input Vu and whatever may be the initial state Vx, the following expressions hold:*

$$\lim_{t\to\infty} E(t) = c \tag{7.22}$$

$$\lim_{t\to\infty} \frac{d}{dt}E(t) = 0 \tag{7.23}$$

where c is a constant value.

Corollary 7.1. *When a CNN transient is exhausted, a constant is always obtained as output, i.e.:*

$$\lim_{t\to\infty} Vyij(t) = c \tag{7.24}$$

or

$$\lim_{t\to\infty} \frac{d}{dt}Vyij(t) = 0 \tag{7.25}$$

where $1 \leq i \leq M$ *and* $1 \leq j \leq N$

From Theorem 7.3 it follows that the state of a cell, while $t \to \infty$ and $dE(t)/dt = 0$, can satisfy, only one of the following three possibilities:

$$\frac{d}{dt}Vxij(t) = 0 \qquad |Vxij(t)| < 1 \tag{7.26}$$

$$\frac{d}{dt}Vxij(t) = 0 \qquad |Vxij(t)| \geq 1 \tag{7.27}$$

$$\frac{d}{dt}Vxij \neq 0 \qquad |Vxij(t)| \geq 1 \tag{7.28}$$

From an examination of the characteristic reported in Figure 7.4 in the portion where $Vyij = f(Vxij)$, (range -1,+1), and from that of (7.19) it can be understood how we arrive at Equations 7.26 to 7.28.

In fact, given the validity of (7.20) when $|Vxij(t)| < 1$, from what was shown in Figure 7.4, $Vyij(t) = Vxij(t)$ and consequently the corresponding time_derivatives will be equal, *i.e.*, $(dVyij(t)/dt) = (dVxij(t)/dt)$: from Theorem 7.3 and its corollary, expression (7.26) follows.

With regard to (7.27), again with reference to Figure 7.4, for $|Vxij(t)| > 1$, $Vyij(t)$ and $Vxij(t)$ are different: $Vyij$ will have the value of +1 or -1, *i.e.* in any case it will be constant hence with zero time derivative; for Expression 7.27 we will have $dVxij(t)/dt = 0$, which is (7.27). Finally, the case of (7.28) is valid for periodic or aperiodic, but limited, $Vxij(t)$ functions.

Now, we will look for the circuit parameter condition for which we can verify only Relation 7.27 for every $Vxij$ in the stable state.

Let us give a stable state definition of a CNN cell. A stable state definition of a CNN cell, V^*xij, is a state variable $Vxij$ of the cell C($i,j,$) satisfying the following conditions:

$$\frac{d}{dt}Vxij(t)\Big|_{Vxij=V^*xij} = 0 \qquad |V^*xij| > 1 \tag{7.29}$$

assuming that: $Vykl = \pm 1$ and $\forall C(k, l) \in Nr(i, j)$.

The stable state of every cell in free evolution depends, therefore, on its initial state and that of the cells bordering on it. A cell can reach a precise stable state if its initial conditions are not varied.

It should be noted that the above definition holds for any combination whatsoever of $Vykl$=1 or $Vykl$=-1 and therefore it may not represent a stable state component of the whole circuit: that means that the individual stability of all the cells in a CNN does not guarantee the CNN stability.

A CNN system *steady state point* is defined as being a state vector having as its components the stable state of the cells: the CNN must always converge on one of the steady state points of the system, when the transient is exhausted.

Every stable state of a CNN steady state system (as just defined above) is an extreme of a set of trajectories of the corresponding differential equations (7.8); this extreme represents a point of confluence, which is to say a union of all the trajectories converging towards on that point. The state space of a CNN can thus be partitioned into a set of trajectories centered on the stable state points of the system.

The following theorem holds:

Theorem 7.4. *If the circuit parameters satisfy the following condition:*

$$A(i,j;i,j) > \frac{1}{Rx} \tag{7.30}$$

then every cell of the CNN will tend, once the transient is exhausted, to a stable state.
Furthermore, the following will then hold:

$$\lim_{t\to\infty} \left|Vxij(t)\right| > 1 \tag{7.31}$$

and

$$\lim_{t\to\infty} Vyij(t) = \pm 1 \tag{7.32}$$

with $1 \le i \le M$ *and* $1 \le j \le N$

From this theorem it can be seen that the outputs converge on two possible values, +1 and -1, and that the circuit will not become chaotic and will not be subject to oscillations

Furthermore, since A(*i,j;i,j*) corresponding to the positive feedback of the cell C(*i,j*), Expression 7.30 sets a lower bound (minimum positive feedback able to guarantee the steady state; it is therefore an indication of the choice of circuit parameters.

7.4 Behavior of a Cellular Neural Network Circuit

Mathematically speaking, a CNN operates as a function having a domain of input values and a codomain of output values. Let us now analyze the circuit in Figure 7.3 of a cell in order to understand its functioning both from the standpoint of the individual basic unit (the cell) and from that of the whole system.

All the elements of the system, including the driven generators, are linear, except for the *Iyx* current generator. The voltage *Vxij* of the cell C(*i,j*) is called state of the cell and its initial condition is presumed to be less than or equal to 1.

The voltage *Vuij* forms the input of the cell C(*i,j*) and is presumed to be a constant of magnitude less than or equal to 1. The voltage *Vyij* is presumed to be the output of the cell C(*i,j*). Every cell contains an independent voltage generator *Eij*, two resistors and at the most *2m* linear voltage driven current generators (with *m* being equal to the number of bordering cells) with controlled voltage, coupled with the bordering cells by means of the control input *Vykl* of every *C*(*k,l*) cell.

An input value of a CNN can therefore be interpreted from the circuit standpoint as a tension: thus with reference to the individual generic cell C(*i,j*), this value will be precisely the *Vuij*. What is interesting is therefore, given a cell, to obtain a system of non-linear differential equations that will represent the state.

This system is given by Expression 7.8 in which the contribution of the circuits of the cells in the neighborhood Nr of the cell in question is supplied, as stated above, by the driven generator currents, connected by linear relationships with the control voltage *Vukl* and *Vykl*, which are, respectively, the input and output of the generic cell *C(k,l)* of the neighborhood; how this control can be exercised, will be seen in detail in the first paragraph of the following chapter, in which the CNN functioning will be correlated with the image processing applications.

The output of the generic cell is characterized by a non-linear relationship:

$$Iyx = \frac{1}{Ry} * f(Vxij) \qquad (7.33)$$

where *f* is the piece wise linear function represented in Figure 7.4; as can be noted, the non-linear current depends directly on the state of the cell considered (*C(i,j)*).

The non-linear outputs *Vykl* of the neighboring cells, multiplied by the corresponding coefficients A will be the cause of the non-linearity of the differential equations (feedback action).

We will see below which are the most suitable methods for solving the differential equation systems representing the CNN state.

The overall state of the CNN will be given by the totality of the cell states, interconnected one with another in the way we have just seen.

7.5 Applications of Cellular Neural Networks to Image Processing

It is well known that image processing in real time is a difficult computational process and that classical processing, segmentation, and interpretation operations are closely conditioned by the type of processor used.

Two problems, the analog-to-digital conversion of the image and the sequential processing of such information, are often the cause of delays in visualization which are unacceptable for real time applications in robotics and monitoring.

Given their ability to perform parallel processing and their high-speed execution, cellular neural networks represent a valid alternative to classical processing methods.

Numerous hardware implementations of cellular neural networks have been proposed [16], [17], [18], and some of them will be integrated in commercial products within a few years.

In [16], a 20*20 CNN was developed, implemented in a $0.7 \mu m$ CMOS technology with a cell time constant equal to $\tau = 5\ \mu s$, which means *ca* 10,000 operations per second, with a global convergence time of *ca* 20 τ.

With a view to showing CNN image processing capacity, we will report below some examples, which, although developed by means of simulation, can be easily implemented on chip.

The steps followed by the simulator during the processing phases are reported in the block diagram of Figure 7.6.

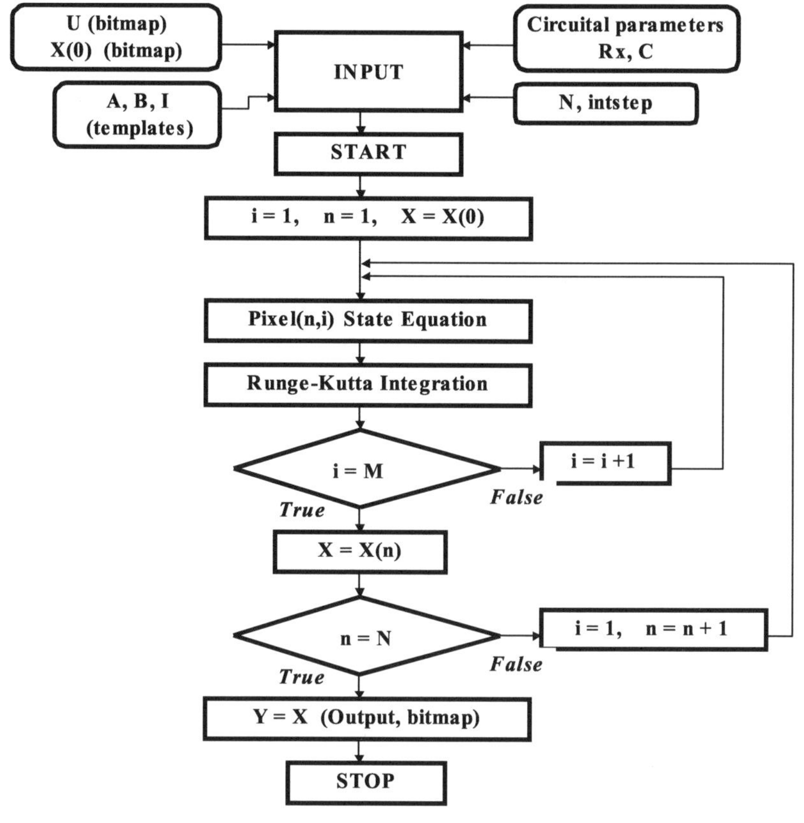

Figure 7.6. Simulator flow chart

7.5.1 Static Image Applications

The two examples reported refer to the operation of *contour extraction*. In the first case on the original image, Figure 7.7a (100*100 pixels, 256 tones of gray) a so-called *Contour* template was applied having the following values: A(1,1)= A(1,2)= A(1,3)= A(2,1)= A(2,3)= A(3,1)= A(3,2)= A(3,3)=0 A(2,2)=2, B(*i,j*)=0, I=0.7 and non-linearity of the form:

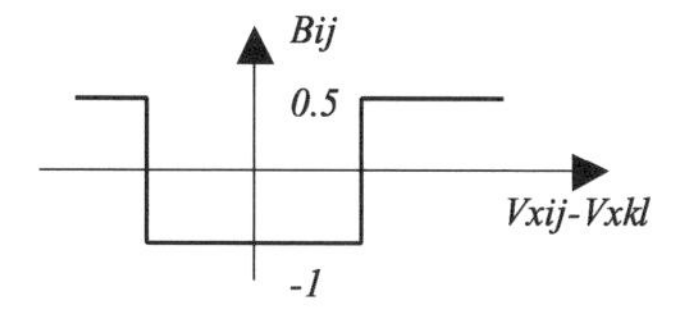

The simulation parameters are the following: *N = 15, Instep = 0.1, Rx = 1.0, C = 1.0* and the final result is reported in Figure 7.7b. Note that, in the final image, the nationality of the airplane is far more legible than in the original.

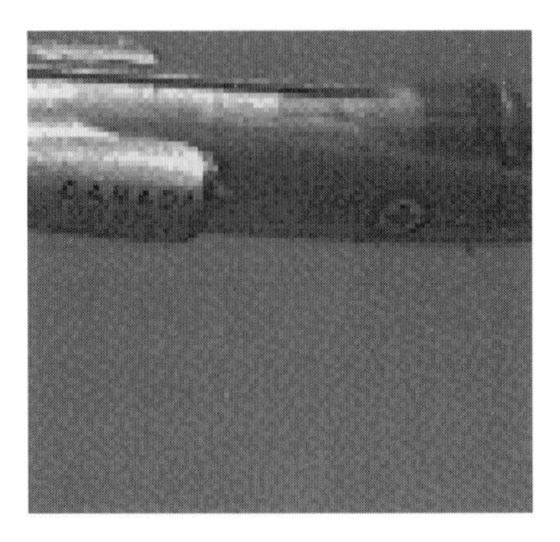

a

b

Figure 7.7. a. original image and b. filtered image

A second application concerns the *contour extraction* of an image of the Mt. Etna volcano during an eruption.

The purpose of the processing was to extract the contours of the smoke emitted by the volcano during the eruptive activity.

With this in view, the three templates used in cascade were THRESHOLD (•), EDGE (•), and DILATA (•) [22] and the parameters used by the simulator were analogous to those of the previous example. The original and filtered images are reported in Figure 7.8a and 7.8b, respectively.

a

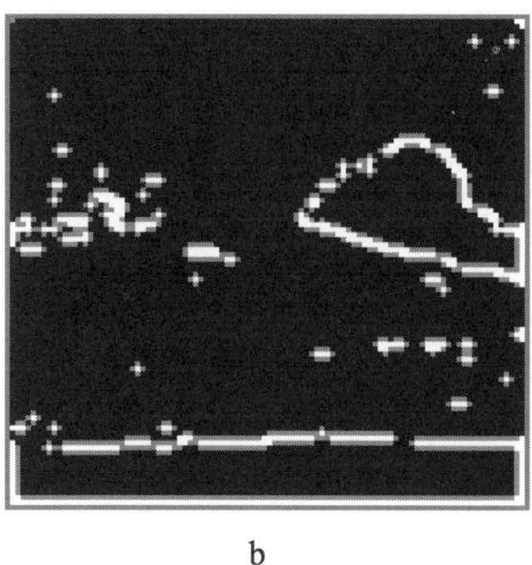

b

Figure 7.8. a. original image and b. filtered image

7.5.2 Automatic Generation of Templates

One of the commonest problems arising in the use of CNNs is the choice of *suitable templates* for the type of processing to be performed.

The literature contains a vast *template library*, but generally a trial and error approach should be used to determine the best template and hence the filter desired.

When using the soft computing paradigm, one uses genetic algorithms for an optimal choice of template, once a filtering operation has been set.

The genetic algorithm fitness function has as its input the original image, and as its target the image already filtered in the way we desire for calculating the phenotype error.

The steps followed by the simulator, in this case integrated with a genetic algorithm, are:

1. loading the original image;
2. loading the final image, that is to say the target one;

3. randomly generating an original template population;
4. assessing the fitness of each element of the population, using both the original and target images;
5. iterating the genetic algorithm until a test condition is verified (maximum number of generations or minimum error reached).

The following fitness function is used:

$$Fitness\,(i) = \sqrt{\sum_{No.pixel} \left| Y_{current} - Y_{target} \right|}^{\,-1}$$

but it can clearly be different and adapted to specific requirements.

By way of example, the template for the inversion was calculated, *i.e.*, the matrices which, given an image, determine its negative. The two input (a) and target (b) images used are reported in Figure 7.9.

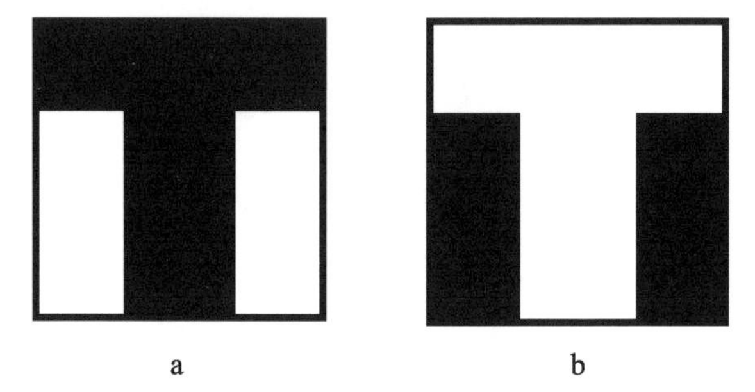

Figure 7.9. a. original image and b. target image

The genetic algorithm parameters are the following:

- population size 110;
- template size 3×3;
- iteration number 200;
- crossover probability 0.65;
- mutation probability 0.004;
- fitness scaling 2.0.

While the simulator parameters are:

- $N = 7$;
- $instep = 0.1$;
- $Rx = 1.0$;
- $C = 1.0$.

The templates and the bias values obtained are the following:

A(1,1)=2.81, A(1,2)=A(2,1)=-2.21, A(1,3)=A(3,1)=2.77,
A(2,3)=-2.53, A(2,3)=A(3,2)=3.91, A(3,3)=-3.77;
B(1,1)=1.82, B(1,2)=B(2,1)=3.9, B(1,3)=B(3,1)=3.11
B(2,2)=-1.77, B(2,3)=B(3,2)=-4.68, B(3,4)=2.09;

I=0.0085

Subsequently, these templates were applied to another image and the results obtained, shown in Figure 7.10, are interesting because the inversion operation was carried out perfectly and the processing times are lower than those measured using the existing templates. In the procedure proposed, the search for a template is conditioned by the image used during the optimization. In order not to generate a *dedicated* filter, that is to say one valid only for the image in question, several images must be presented to the genetic algorithm so as to generalize the CNN's ability to filter.

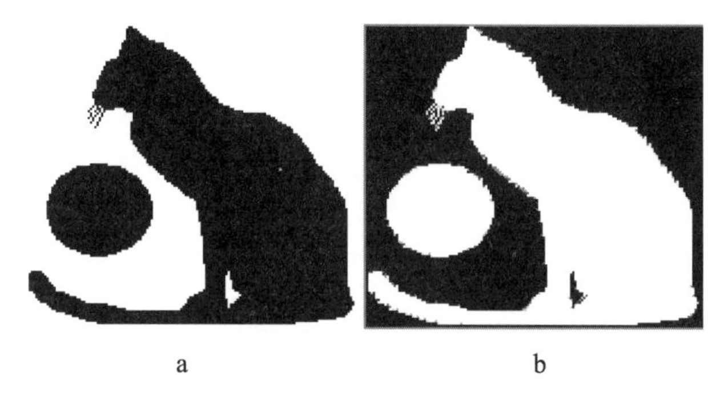

a b

Figure 7.10. a. original image and b. filtered images

7.5.3 Enlargement and Reconstruction of Images

The procedure described consists in *zooming* and then *filtering* the original image so as to allow it to maintain a high quality enlarged image. The possible industrial applications of this procedure range from digital video cameras to scanners and to fax transmissions.

The steps taken by the simulator are:

1. loading the original image;
2. doubling the size of the original image and adding white pixels to the original ones;
3. reconstructing the missing pixels by means of the *Zoom_1* and *Zoom_2* templates [22] reported below:
4. putting the enlarged image into focus by use of the *Focus_1* and *Focus_2* templates [22].

In particular, after Step 2 in the enlarged image, the missing pixels are reconstructed as shown in Figure 7.11 following the reported steps.

The *M* pixels are determined by interpolating the neighboring pixels of the original image.

The *X* pixels are determined by using the *Zoom_1* template [22] on the image which has been widened at Step 2.

The *Y* and *T* pixel are determinated by using the *Zoom-2* template on the image which has been widened at Step 2.

At the end, the *Focus_1* and *Focus_2* templates are applied to improve the quality of the image obtained. The procedure described was applied to the image reported in Figure 7.12 and the results are encouraging because they are better than those obtained using standard procedures.

The fields of application mentioned in this chapter are only part of the numerous CNN applications.

In fact, CNNs were born in the *electronic* world and their first applications were concerned exclusively with *image processing*.

Their basic characteristics, non-linearity and modularity, have subsequently contributed to their being massively used in fields as various as, for example, solving partial differential equations, generating dynamic complexes, generating chaotic signals for cryptography, and creating autowaves for locomotion.

1	2	3	4	5
6	7	8	9	10
11	12	13	14	15
16	17	18	19	20

1	M	2	M	3	M	4	M	5
M	X	Y	X	Y	X	Y	X	M
6	T	7	T	8	T	9	T	10
M	X	Y	X	Y	X	Y	X	M
11	T	12	T	13	T	14	T	15
M	X	Y	X	Y	X	Y	X	M
16	M	17	M	18	M	19	M	20

Figure 7.11. Reconstruction of the image

Figure 7.12. Applications of zooming and filtering

7.6 References

1. Chua LO .CNN: A Paradigm For Complexity. In World Scientific. 1998
2. Arena P, Baglio S, Fortuna L, Manganaro G. Cellular Neural Networks: A survey. IFAC Symp. on Large Scale Systems, London, UK, pp. 53-58, vol. 1, 1995
3. Arena P, Fortuna L, Manganaro G, Spina S. CNN Image Processing for the Automatic Classification of Oranges. In CNNA-94 Third IEEE International Workshop on Cellular Neural Networks and their Applications, Rome, 1994
4. Cardarilli GC, Sargeni F. Very Efficient VLSI Implementation of CNN with Discrete Templates. In IEE Electronics Letters. 1993; 29: 14, 1286-87
5. Salerno M, Sargeni F, Bonaiuto V. 6×6 DPCNN: A Programmable Mixed Analogue-Digital Chip for Cellular Neural Networks. In 4th IEEE Int. Workshop on Cellular Neural Networks and their Appls, Seville. 1996; 451-456
6. Roska T, Wu CW, Balsi M, Chua LO. Stability and Dynamics of Delay-Type General and Cellular Neural Networks In IEEE Transactions on Circuits and Systems-I: Fundamental Theory and Applications. June 1992; 39, 6
7. Chua LO , Roska T. The CNN Paradigm. In IEEE Transaction on Circuits and Systems. March 1993; I: 40: 147-156
8. Special Issue on Cellular Neural Networks.In International Journal on Circuit Theory and Applications. September-October 1992; 20: 5
9. Special Issue on Cellular Neural Networks. In IEEE Transactions on Circuit and Systems. March 1993; 40
10. Special Issue on Cellular Neural Networks.In International Journal on Circuit Theory and Applications. 1996; 24: 1-3
11. Chua LO, Wu CW. The Universe of Stable CNN Templates. In International Journal of Circuit Theory and Applications. July-August 1992; 20.497-517
12. Arena P, Baglio S, Fortuna L, Manganaro G. State Controlled CNN: A New Strategy for Generating High Complex Dynamics. In IEEE Trans. on Circuits and Systems - PartI
13. Arena P, Baglio S, Fortuna L, Manganaro G. Chua's Ciurcuit can be Generated by CNN Cells.In IEEE Trans. on Circuits and Systems. February 1995; I: 42: 2
14. Arena P, Baglio S, Fortuna L, Manganaro G. Generation of N-Double Scroll Via Cellular Neural Networks. In International Journal on Circuit Theory and Applications. 1996; 24: 241-252
15. Chua LO. Global Unfolding of Chua's Circuit. In IEICE Trans. on Fundamentals. 1993; E76-A: 704-734
16. Espejo S, Rodriguez-Vasquez A, Dominguez-Castro R, Huertas J, Sanchez-Sinencio E. Smart-pixel cellular neural networks in analog current mode CMOS technology. In J. of Solid-State Circuits. 1994; 29.895-905
17. Kinget P, Steyaert M. An Analog Parallel Array Processor for Real Time Sensor Signal Processing. In Digest of Tech. Papers IEEE Int. Solid-State Circuits Conf. (ISSCC'96). 1996; 92-93
18. Espejo S, Carmona R, Dominguez-Castro R, Rodriguez-Vasquez A. A CNN universal chip in CMOS technology. In Int. J. of Circuit Theory and Applications. 1996; 24. 93-109
19. Arena P, Baglio S, Fortuna L, Manganaro G. Hyperchaos from Cellular Neural Networks. In IEE Electronic Letters. February 1995; 31: 4: 250-251
20. Arena P, Caponetto R, Fortuna L, Manganaro G. Cellular Neural Networks to Explore Complexity. In Soft Computing Research Journal. September 1997; 1: 3: 120-136

21. Manganaro G, Arena P, Fortuna L. Cellular Neural Networks: Chaos, Complexity and VLSI processing. Berlin: Springer-Verlag, 1999
22. Caponetto R, Lavorgna M, Matinez A, Occhipinti L. Cellular Neural Network Simulator for Image Processing. CNNA. London UK, 14-17 April 1998

8. Complex Dynamics and Cellular Neural Networks

8.1 Introduction

Artificial neural networks, fuzzy systems, and cellular neural networks are non-linear systems. Furthermore, their use is directed towards the modeling and control of dynamic non-linear systems. Soft computing techniques would, in any case, be superfluous for studying linear systems, being more often than not dedicated to the study of systems having a certain degree of complexity. Hence the need to introduce the basic tools for analyzing complex dynamic systems. This chapter is therefore arranged around two themes: the first regards the analysis of complex dynamic systems, while the second deals with the processing of an innovative procedure for generating complex dynamics by means of CNNs and thus by means of soft computing types of techniques.

8.2 Notes on Dynamical Systems

Let us take a *dynamic system* of finite size, regular, with state transition function, and output transformation η. We then have:

$$\begin{aligned} x(t) &= \varphi(t_0, t, x_0, u(\cdot)) \\ y(t) &= \eta(t, x(t_1), u(\cdot)) \end{aligned} \tag{8.1}$$

where:

$x \in \mathrm{X}$ is the state;
$y \in \Gamma$ is the output;
$t_0 \in \mathrm{T}$ is the initial instant;
$x_0 \in \mathrm{X}$ is the initial state;
$u(\cdot) \in \Omega$ is the input.

It is well known that the *movement* $x(\cdot) = \varphi(t_0, \cdot, x_0, u(\cdot))$ of a regular system of finite size is the solution of a vector differential equation of the type:

$$\frac{dx(t)}{dt} = f(x(t), u(t), t) \tag{8.2}$$

with an initial condition $x(t_0) = x_0$. The *state transition function* φ represents the solution of Equation 8.2.

Dynamic systems can be distinguished into *autonomous* and *non-autonomous* according to whether the state transition function depends or does not depend on an external input; autonomous systems are indicated for short as:

$$\dot{x} = f(x)$$

whereas non-autonomous ones are explicitly considered functions of time:

$$\dot{x} = f(x, t)$$

Indicating for short with $\varphi_t(x_0)$ the state transition function, the set of points of the state space is called the *trajectory for* x_0.

Definition 8.1. *Given the dynamic system (8.2), the flow* $\varphi_t : R^n \to R^n$ *is the set of all the solutions of (8.2) in X.*

For example, in the case of a linear dynamic system defined by the equation:

$$\dot{x} = Ax$$

with $A \in R^{n \times n}$ e $x \in X \subseteq R^n$, the state transition function will be:

$$\varphi_t(x_0) = e^{At} x_0$$

The operator e^{At} therefore contains information regarding the set of all the solutions and it can be stated that e^{At} defines a *flow* in R^n and that such a flow is generated by the function $f(x) = Ax$.

If a value $\tau > 0$ exists such that $f(x, t) = f(x, t+\tau)$ for every value of x, the system is known as periodic of the period τ. All the non-autonomous dynamic systems considered below will be presumed to be *periodic* and τ will be the minimum period. A non-autonomous dynamic system of the order n of the type described by (8.2) can be converted into an autonomous one of the order n+1 by means of a simple change of variables.

In fact, supposing that $\theta = 2\pi t / \tau$, the following system is obtained:

$$\dot{x} = f(x, \theta\tau / 2\pi) x(t_0) = x_0 \tag{8.3}$$

$$\dot{\theta} = 2\pi / \tau \theta(t_0) = 2\pi t_0 / \tau \tag{8.4}$$

Since f is periodic with period τ, the system of equations (8.3), (8.4) will be periodic with period 2τ; considering, therefore, the planes $\theta = 0$ and $\theta = 2\pi$, the state space can be transformed by the Euclidean one R^{n+1} to the cylindrical space $R^n \times S^1$, where $S^1 = [0,2\pi]$ denotes the circle associated with the periodic input function.

The solution of the system of equations (8.3) and (8.4) in the cylindrical space can thus be expressed as:

$$\begin{bmatrix} x(t) \\ \theta(t) \end{bmatrix} = \begin{bmatrix} \varphi_t(x_0) \\ (2\pi t / T) \bmod 2\pi \end{bmatrix} \tag{8.5}$$

and by means of this transformation of variables, the results obtained for the autonomous systems can be applied analogously to the non-autonomous ones. A basic hypothesis when carrying out analyses of dynamic systems is to consider the flow of the system (8.2) a differomorphism.

Definition 8.2. *A map* g: $X \to Y$, *continuous with its first derivative, is a differomorphism if the inverse map* $g^{-1}: Y \to X$ *is also continuous with its first derivative.*

This hypothesis is not restrictive and implies the following [1]:

- the solution of the dynamic system (8.2) exists for all values of t;
- the trajectory of an autonomous system is univocally specified by the initial conditions, and thus two trajectories emerging from two distinct points of the state space cannot intersect;
- with respect to the initial conditions, the derivative of the trajectory exists and is not individual.

The term regime behavior is referred to the trend of the magnitudes of the system evaluated for $t \to \infty$. When the transient is exhausted, the motion of the trajectory $x(t)$ occurs in a region of the state space known as the attractor; in particular, in the case of dissipation systems, the volume occupied by the attractor is generally small with respect to that by the state space.

Definition 8.3. *A system is a dissipative one if the volume in which the state trajectories are contained is contracted by the application of the flow [2].*

With the regime analysis of dynamic systems, the definition of the *limit set* assumes great importance. Let us consider the dynamic system defined by the following equations:

$$\dot{x} = f(x,t), \quad with \quad x(0) = x_0 \tag{8.6}$$

and let $X \subset R^n$ be the state space.

Definition 8.4. *Considering a point $x \in X$ of the state space, a point $y \in X$ is an ω-limit point of x if, for every neighborhood $I(y) \subset X$ of y, the state transition function $\varphi_t(x)$ repeatedly falls within $I(y)$ for* $t \to \infty$.

Definition 8.5. *The set $L(x)$ of all the ω-limit points of x is known as the ω-limit set of x.*

Definition 8.6. *An ω-limit set L(x) is attractive if there exists an open neighborhood E of L(x) such that the ω-limit set of every point z in E is precisely L(x). The union of all the sets E for which an ω-limit set is attractive is known as the basin of attraction.*

In other words, B_L is the set of all the initial conditions which tend towards $L(x)$ for $t \to \infty$.

Below, we will briefly refer to a limit set rather than to a *ω-limit set*; the attractive limit sets are the only ones of interest in that they are the only ones that can be observed experimentally or in simulation. These definitions can easily be extended to autonomous systems after effecting the transformation (8.5), with reference to the autonomous system defined in the cylindrical state space.

For a linear dynamic system there exists one limit set only, *i.e.*, one regime operation condition only, and the basin of attraction is the whole state space. These statements are no longer valid for non-linear dynamic systems in which different attractive limit sets are generally present, to each of which a different basin of attraction is associated.

An *equilibrium state* x_{eq} of an autonomous dynamic system is a constant solution of Equation 8.2, and thus we have:

$$\varphi_t\left(x_{eq}\right) = x_{eq} \tag{8.7}$$

In correspondence with x_{eq}, the function $f(x)$ becomes zero, and therefore the equilibrium states are obtained from the solution of the equation $f(x) = 0$.

The limit set of an equilibrium state is the state itself, *i.e.*, the point x_e of the state space, whereas in the frequency domain, the spectrum is made up only of a peak at the origin.

Let us consider, for example, the following equations with $\varepsilon = 0.4$:

$$\dot{x} = y \tag{8.8}$$

$$\dot{y} = -\varepsilon\, y - \sin\left(x\right) \tag{8.9}$$

which represent the motion of a damped non-linear pendulum [1].

It can readily be noted that this system possesses infinite equilibrium points $(x, y)=(k\pi ,0)$ per $k=0, \pm 1, \pm 2,\ldots$; of these, those corresponding to odd values of k are attractive as shown in the state trajectory reported in Figure 8.1.

A *periodic solution* of the system (8.2) is expressed by the following relation:

$$\varphi_t\left(x^*\right) = \varphi_{t+\tau}\left(x^*\right) \tag{8.10}$$

where τ represents the minimum period of the solution. It can be observed that the above-mentioned relation is valid not only for x^*, but for all the state values belonging to the periodic solution.

If it is not possible to determine a neighborhood of solution (8.10) containing other periodic solutions, this is called *isolated*; for an autonomous system, a solution of this kind is called a *limit cycle.*

The limit set of a limit cycle is the closed loop followed by $\varphi_t(x^*)$ in a period, which is a differomorphic copy of a circle [1].

In the case of non-autonomous systems, the *periodic solution* is expressed as a function of the initial time t_0:

$$\varphi_t\left(x^*, t_0\right) = \varphi_{t+\tau}\left(x^*, t_0\right) \tag{8.11}$$

Transforming the non-autonomous system into an autonomous one by means of (8.3) and (8.4), a limit cycle is obtained in the cylindrical state space. If the period τ_f of the forcing signals a submultiple according to k of the minimum period τ, solution (8.11) is called *k-periodic*. The spectrum of a periodic solution is made up of a series of peaks, each one centered on a multiple of the basic frequency. Figure 8.2 reports the limit cycle of the *Van der Pol oscillator*, which is a second-order autonomous system defined by the following equations [3]:

$$\begin{aligned} \dot{x} &= y \\ \dot{y} &= \left(1 - x^2\right) + x \end{aligned} \tag{8.12}$$

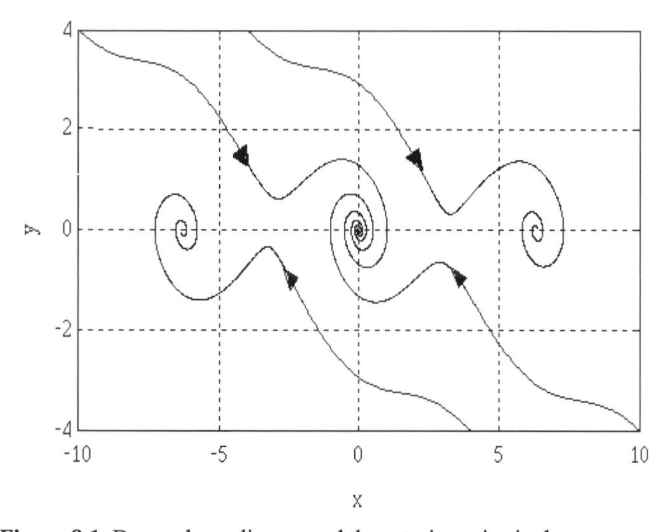

Figure 8.1. Damped non-linear pendulum: trajectories in the state space

The case of a *k-periodic* oscillation is reported in Figure 8.3 with reference to the following equations which model the *Duffing oscillator* [3]:

$$\begin{aligned} \dot{x} &= y \\ \dot{y} &= x - x^3 - \varepsilon y + \cos(\omega t) \end{aligned} \tag{8.13}$$

In this case, $k = 3$ and the parameters have the values $\varepsilon = 0.22$, $\gamma = 0.3$, $\omega = 1.0$. The k-periodic solutions are often called *sub-harmonics*. It is important to note that the presence of a sub-harmonic oscillation makes sense only in the case of non-autonomous dynamic systems; in fact, these are the only ones for which a reference frequency can be defined. One *quasi-period* solution of the system (8.2) can be expressed as the sum of a numerable set of periodic functions:

$$x(t) = \sum_i h_i(t) \tag{8.14}$$

Each of the functions h_i has a minimum period τ_i; furthermore there exists a finite set of linearly independent frequencies, called *basic frequencies*, such that each $f_i = 1/\tau_i$ can be expressed as a linear combination of the frequencies of this set. The limit set of a quasi-periodic solution having m basic frequencies is a differomorphic copy of an m-torus $T^n = S^l \; x \; S^l ... x \; S^l$, where each S^l represents the circle corresponding to one of the basic frequencies. If, for example, there are only two basic frequencies, it can be proved that the trajectory will pass arbitrarily close to each point of a torus [1]; therefore, the torus is the limit set of the system. Figure 8.4 reports the state trajectories of the Van der Pol oscillator with a periodic external forcing, obtained by adding a periodic term with a pulsation of $\omega = 6$ to the second equation of (8.12). For these parameter values, an almost periodic operation is obtained [1]. The spectrum of any dimension whatsoever of the system will be concentrated around certain peaks which, unlike what happens in the *k-periodic* case, will no longer be regularly spaced out. The limit sets seen previously all have in common a somewhat regular structure, whether they be a point, a closed loop or a surface.

The trajectories of some non-linear dynamic systems converge, instead, towards geometric objects in the state space that cannot be classified as some of the attractors considered previously: these attractors display an extremely complex structure.

They are called *strange attractors* and are closely linked to the *Cantor sets* [1] and the *fractal sets*; in fact, they generally have an non integer dimension.

Although the definitions of attractor and limit set are slightly different [2], from an operative standpoint the error committed by not taking this difference into account is a negligible one [1].

The temporal evolution waveform of a dimension of such a system does not display any periodicity and appears stochastic; moreover, the spectrum is no longer formed by more or less distinct peaks, but has a continuous pattern similar to that of white noise.

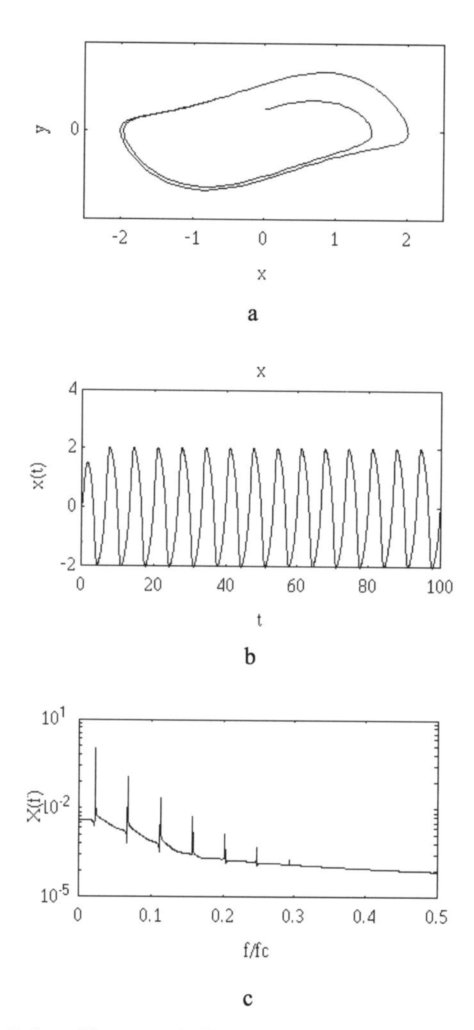

Figure 8.2. Van der Pol oscillator, periodic solution: a. state trajectory; b. waveform of $x(t)$; c. spectrum of $x(t)$

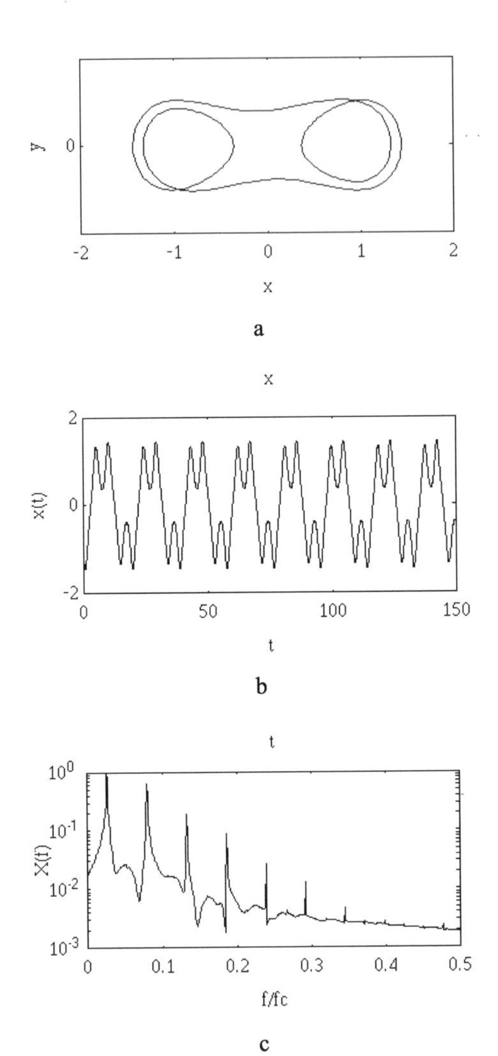

Figure 8.3. Duffing oscillator, *k* periodic solution (k=3): a. state trajectory b. waveform of $x(t)$; c. fourier spectrum of $x(t)$

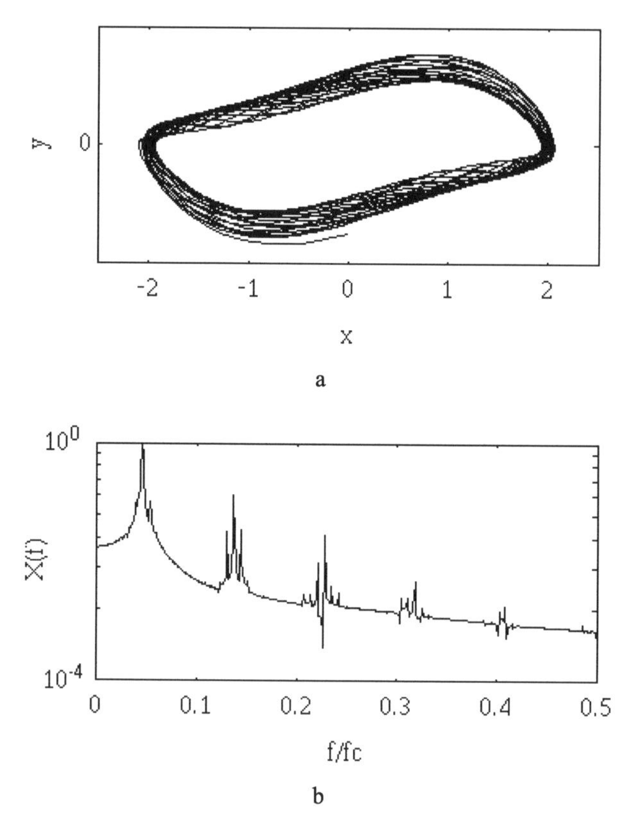

Figure 8.4. Van der Pol oscillator with periodic forcing, quasi-periodic attractor: a. state trajectories; b. spectrum of $x(t)$

One feature common to all dynamic systems that display this type of attractor is their *sensitivity to the initial conditions*: two trajectories emerging from two points at any degree of proximity diverge exponentially in time.

For *dissipative systems*, the attractor must be contained in a finite volume of the state space, while the contradiction between the diverging behavior of the proximate trajectories and the finite volume in which the attractor is contained cannot be solved by the concepts of Euclidean geometry, but instead by *fractal geometry theory*.

One of the best-known dynamic systems whose regime trajectories are based on a strange attractor is that described by the following non-linear differential equations, the so-called *Lorenz equations* [4]:

$$\begin{aligned} \dot{x} &= \sigma(y - x) \\ \dot{y} &= -xz + rx - y \\ \dot{z} &= xy - bz \end{aligned} \tag{8.15}$$

Eduard Lorenz developed these equations for studying the behavior of atmospheric phenomena on the basis of an analogy with the convective motion of a fluid contained in an closed volume and being subject to difference in temperature. The parameters σ, r, and b are proportional, respectively, to the number of Prandtl, to the number of Rayleigh, and to the physical dimensions of the volume containing the fluid, their nominal values being $\sigma = 10$, $r = 8/3$, and $b = 28$. The state variables x, y, and z represent, respectively, the intensity of the convective motion, the difference in temperature between the ascending and the descending current, and the distortion of the temperature vertical profile.

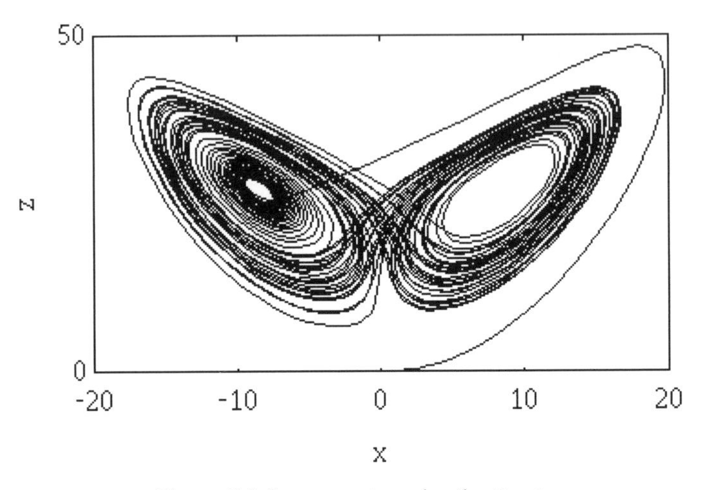

Figure 8.5. Lorenz system chaotic attractor

The attractor of System 8.15 is reported in Figure 8.5. The flow of this system contracts the volume in all the state space [5], and so the volume of the attractor is zero in R^3; therefore, if the attractor is not a point of equilibrium or a limit cycle, it will be a strange attractor.

8.3. Chaotic Systems

The analysis of non-linear circuits and systems has led to the discovery of a particularly strange phenomenon in their regime behavior: *chaos* [6]. Often the study of chaos in non-linear dynamic systems has been carried out in a prevalently theoretical way, and at times almost speculatively [7-8]: what is more, in the course of the last years scientific research in this field has markedly developed and revealed a high degree of order in some behaviors which were previously considered to be random [5]; chaos has thus been identified as an original theme, closely connected with concepts peculiar to the theory of circuits and to that of systems.

Interest shown by the engineering profession in chaotic phenomena has considerably grown, mainly due to the numerous results obtained by scientific research regarding increasingly new and more varied applications [9-10], including the field of industry [11], which exploit the peculiar characteristics of chaotic systems and the huge amount of information contained in their regime behavior [12].

A *chaotic system* can be synthetically defined as a deterministic system for which it is not possible to predict the regime condition.

The apparent contradiction in terms contained in this statement is justified by the fact that this kind of systems displays so high a sensitivity to the initial conditions that two trajectories emerging from two infinitely proximate points will evolve in a totally different way, although they will both stay confined in an enclosed region of the state space.

Thus, since the value of the initial conditions cannot be assigned with infinite precision, and since generally we are unable to determine the state transition function in its enclosed form, we are unable to predict the regime value of the system due to its high level of sensitivity towards the initial conditions; and that despite the fact that we know the differential equations governing the temporal evolution of the system.

A more formal definition of a chaotic system is that based on the concepts of density, transitivity, and sensitivity, which will be outlined below [13].

Definition 8.7. *Let us consider a vector space X in which a metric d has been defined. A subset* $B \subseteq X$ *is said to be dense in X if the set of accumulation points of B is equal to X. A succession* $\{x_n\}^{\infty}{}_{n=0}$ *of points in X is dense in X if for every point y of X there exists a sub-sequence* $\{x_n\}^{\infty}{}_{n=0}$ *which converges on y. An orbit* $\{x_n\}^{\infty}{}_{n=0}$ *of a dynamic system f defined in the space X is dense in X if the succession of points* $\{x_n\}^{\infty}{}_{n=0}$ *is dense in X.*

Definition 8.8. *A dynamic system:*

$$\dot{x} = f(x)$$

is said to be transitive if, considering any two open subsets of X, A *and* B, *there exists a finite value of* n *for which we have:*

$$A \cap f^{on}(B) \neq \varnothing$$

Definition 8.9. *Let us consider a dynamic system:*

$$\dot{x} = f(x)$$

defined in a state space X, and let d *be the metric defined in X.*

A dynamic system is said to be sensitive to the initial conditions if there exists a number $\delta > 0$ *such that for every x in X and for every neighborhood B of x of radius* $\varepsilon > 0$*, there exists a point* $y \in B$ *and a neighborhood* $n > 0$ *for which* $d(f^{on}(x), f^{on}(y)) > \delta$.

The operator f^{on} is a transformation $f^{on}: X \rightarrow X$ which has the following properties: $f^{o0}(x) = x$; $f^{o1}(x) = f(x)$; $f^{on}(^{x}) = f(f^{o(n-1)}(x))$. It can be observed that in a transitive system there is a kind of mixing of the trajectories in the state space which allows conciliation of the apparently antithetical positions of the previously seen neighboring orbits exponential divergence and of the finite attractor volume. It is thus possible to introduce a chaotic system definition [13].

Definition 8.10. *A dynamic system*

$$\dot{x} = f(x)$$

defined in the state space $X \subseteq R^n$ is chaotic if:

- *it is transitive;*
- *it is sensitive to the initial conditions;*
- *the periodic solutions of f are dense in the state space.*

The "mixing" of the regime trajectories of a chaotic system occurs in the "fine" structure of strange attractors.

Definition 8.11. *From an operative standpoint, an attractor can be defined as an attractive limit set that contains dense orbits [14].*

One strategy for the qualitative analysis of dynamic systems has been given by the French mathematician Poincaré who substituted the flow of a system of the order n with a discreet system of the order n-1, the so-called *Poincaré map*. The analysis of such a map facilitates the study of non-linear dynamic system attractors in so far that it allows one to work with a system of a lesser order, and moreover evidences the structure of some attractors that appear to be particularly intricate in the state space.

For a non-autonomous system having a minimum period τ, the Poincaré map is defined as the sampling, at the minimum period, of the flow $\varphi_t(x(t))$. In fact, by means of (8.3) and (8.4), a non-autonomous system of the order n can be transformed into an autonomous system of the order n+1 defined in the cylindrical space $R^n \times S^1$. The Poincaré map P_N for this type of system is formally defined as the intersection of the flow with the hyperplane:

$$\sum := \left\{ (x,\theta) \in R^n \times S^1 : \theta = \theta_0 \right\}$$

Therefore, the map P_N: $\Sigma \rightarrow \Sigma$ will be:

$$P_N(x) := \varphi_{t_0+\tau}(x, t_0) \tag{8.16}$$

In the case of autonomous systems, the Poincaré map P_A is defined only locally in the neighborhood of a point $x^* \in X$, which represents the value of the state of the system in the event of a periodic solution with minimum period T, as the intersection of the trajectory emerging from x^* with the hyperplane Γ transversal to the flow in x^*.

It can be observed that for autonomous systems we cannot ensure that a trajectory emerging from a generic point Γ will again intersect the hyperplane; hence the need for the map to be locally defined.

Let us consider, then, a non-autonomous dynamic system with minimum period τ described by the following equations:

$$\dot{x} = f(x,t) \tag{8.17}$$

and let us analyze the representation of the various attractors in the Poincaré map.

The Poincaré map of a *k-periodic* attractor will be formed by k points in the hyperplane Σ, while a quasi-periodic attractor will be composed of a set of points all belonging to the same loop. The Poincaré map of a chaotic attractor will instead have a *fine structure* formed of a set of points that describe an object in the Poincaré hyperplane.

Figure 8.6 shows the Poincaré map of the Duffing oscillator chaotic attractor and the corresponding state trajectories; the equations describing the motion of the system are those of (8.13) with the parameter values $\varepsilon = 0.25$, $\gamma = 0.3$, $\omega = 1.0$.

It should be noted how the Poincaré map reveals the presence of a well-defined structure in the attractor, which does not appear, instead, from the analysis of the state trajectories.

Below, we will introduce some quantities that will allow us to characterize the quality of the regime conditions.

Intuitively one can say that an attractor L is stable if all the trajectories emerging from a neighborhood of L remain in the neighborhood of the attractor.

Definition 8.12. *A limit set L is stable if, for every open neighborhood U of L, there exists an open neighborhood V of L such that for every* $x \in V$ *and for every* $t > 0$ *we have:* $\varphi_t(V) \subset U$ *or, equivalently,* $\varphi_t(V) \subset U$ *for every value of* $t > 0$.

If all the trajectories contained in a neighborhood of L are attracted by this, the attractor will be asymptotically stable.

Definition 8.13. *A limit set L is said to be asymptotically stable if there exists an open neighborhood V of L such that the limit set of every point in V is precisely L.*

Conversely, if every trajectory, contained in a neighborhood of the attractor, but different from the one making up the limit set, is rejected, the attractor is said to be unstable.

Definition 8.14. *A limit set L is unstable if there exists an open neighborhood U of L such that the* α*-limit set (the limit set for* $t \to -\infty$*) of every point of U is precisely L.*

Definition 8.15. *A limit set L is stable if every neighborhood V of L contains at least one point whose* α*-limit set is L.*

Let us consider, then, the dynamic system described by the following equations:

$$\dot{x} = f(x,t), \quad with \quad x(0) = x_0 \tag{8.18}$$

If this has an equilibrium state in x_{eq}, the behavior of the system in a neighborhood of the point x_{eq} of the state space can be studied by determining the eigenvalues of the following linear system which describes the evolution in time of a perturbation $\delta x(t)$ of the equilibrium state x_{eq} of the system (8.18):

$$\delta\dot{x} = Df(x_{eq})\delta x \quad \delta(x(0)) = \delta x_0 \tag{8.19}$$

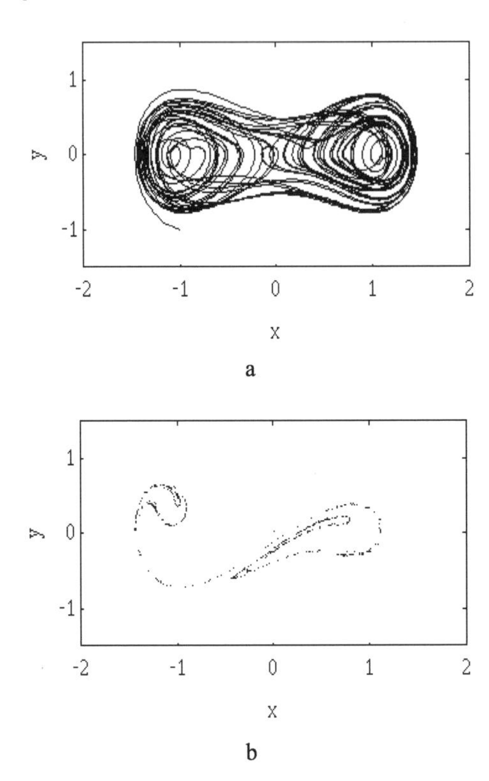

Figure 8.6. Duffing oscillator: a. state trajectories; b. Poincaré map of the chaotic attractor

The stability of the equilibrium state will thus depend on the position of the eigenvalues of $Df(x_{eq})$ in the overall plan.

Let us suppose, for the sake of simplicity, that the eigenvalues are all distinct from the trajectory emerging from $x_{eq} + \delta x_0$, consequently:

$$\begin{aligned} \varphi_t(x_{eq} + \delta x(t)) &= x_{eq} + \delta x(t) = x_{eq} + e^{Df(x_{eq})t}\delta x_0 \\ &= x_{eq} + c_1 e^{\lambda_1 t}\overline{\eta}_1 + ... + c_n e^{\lambda_n t}\overline{\eta}_n \end{aligned} \tag{8.20}$$

where c_1, c_2,....c_n, are constants determined by the initial conditions and η_1, ..., η_n are the eigenvalues corresponding to the eigenvalues λ_1, ...,λ_n. It is thus evident that the eigenvalue λ_i measures the contraction ($\lambda_i<0$) or the expansion ($\lambda_i>0$) of the flow in a neighborhood of x_{eq} in the direction represented by the eigenvalue η_i.

The equilibrium state will then be said to be *asymptotically stable* if all the eigenvalues have a negative real part, and not stable (saddle point) if there are eigenvalues with a real part both positive and negative. In the case where one or more eigenvalues have a zero real part, the stability of the linearized system is no longer correlated with that of the non-linear system equilibrium state [15]; in that case, in order to obtain a characterization of the equilibrium state x_{eq}, recourse must be had to the second criterion of Lyapunov [15].

The stability of the periodic solutions is determined by the characteristic multipliers which are a generalization of the eigenvalues concept. A periodic solution corresponds to a fixed point x^* in the Poincaré map, and thus the stability of the periodic solution may be derived from the Poincaré map.

Considering Relation 8.16 in a neighborhood of the fixed point x^*, the following linear map is obtained:

$$\delta x_{k+1} = DP_N(x^*)\delta x_k \tag{8.21}$$

which describes the evolution of a perturbation of the point x^*.

If we suppose the dimension of the Poincaré map to be equal to p, $p=n$ for non-autonomous systems and $p=n-1$ for autonomous systems, the eigenvalues m_i ($i=1, \ldots, p$) of the matrix $DP_N(x^*)$ are said to be characteristic multipliers of the periodic solution. Analogously to what was seen in the case of the equilibrium state for the eigenvalues, the position of the characteristic multipliers determines the quality of the equilibrium of the fixed point x^*, and hence the stability of the periodic solution.

If $|m_i| < 1$ for every i, the periodic solution will be asymptotically stable: in the state space, the trajectories contained in a neighborhood of the periodic solution will tend to this asymptotically. Analogously, we can define the instability of the solution and the non-stability (saddle point).

8.4 The Lyapunov Exponents

In the previous paragraphs, the flow φ_t was defined as an application between two vector spaces enjoying the composition law $\varphi_{t2+t1} = \varphi_{t2}$ or φ_{t1}; therefore, the action of the flow is that of mapping an initial condition x_0 in $x(t)$, and the trajectory is obtained with variations of t.

Given two points $(x,y) \in X$, belonging to the state space X in which a metric d has been defined, let us consider the following limit:

$$\lim_{d(x,y)\to 0} \frac{d(\varphi_t(x), \varphi_t(y))}{d(x,y)} \tag{8.22}$$

With a view to quantifying this limit, let us consider the curve $C(s)$ such that $C(0) = x$, and thus the limit for $d(x,y)$ which tends to zero can be rewritten as follows:

$$\lim_{s\to 0}\frac{d(\varphi_t(C(s)),\varphi_t(x))}{d(C(s),x)} \tag{8.23}$$

The *Lyapunov exponents,* $\lambda_1, \ldots, \lambda_n$, are defined by the following relation [16]:

$$\lim_{t\to\infty}\frac{1}{t}\ln\lim_{s\to 0}\frac{d(\varphi_t(C(s)),\varphi_t(x))}{d(C(s),x)} \tag{8.24}$$

A definition of greater utility for determining the value of the Lyapunov exponents is obtained by referring to the *variational equation.* Considering the state transition function $\varphi_t(x_0)$, for a dynamic system of the order n, we can write:

$$\dot{\varphi}_t(x_0, y_0) = f(\varphi_t(x_0, y_0), t), \quad \varphi_{t0}(x_0, t_0) = x_0 \tag{8.25}$$

Differentiating the previous relation with respect to the initial condition x_0, we obtain:

$$\begin{aligned} &D_{x_0}\dot{\varphi}_t(x_0,t_0) = D_x f(\varphi_t(x_0,t_0),t) D_{x_0}\varphi_t(x_0,t_0), \\ &D_{x_0}\varphi_{t_o}(x_0,t_0) = I; \end{aligned} \tag{8.26}$$

where $I \in R^{n\times n}$ is the identity matrix. Given $\Phi_t(x_0, t_0) := D_{x0}\varphi_t(x_0, t_0)$, the previous equation becomes:

$$\dot{\Phi}_t(x_0,t_0) = D_x f(\varphi_t(x_0,t_0),t)\Phi_t(x_0,t_0), \quad \Phi_{t_o}(x_0,t_0) = I \tag{8.27}$$

which is said to be a variational equation.

This is a linear differential equation, varying in time, with matrix coefficients. Since the initial condition is the identity matrix, it follows that $\Phi_t(x_0,t_0)$ is the state matrix of the linear dynamic system (8.27); thus a perturbation δx_0 of the initial state x_0 will evolve as follows:

$$\delta x(t) = \dot{\Phi}_t(x_0,t_0)\delta x_0 \tag{8.28}$$

Let us consider now the linearized system in a fixed point neighborhood x^* of the Poincaré map of a non-autonomous system with minimum period τ. In this case, we have:

$$P_N(\dot{x}) = \varphi_\tau(x, t_0)$$

and thus:

$$DP_N(x^*) = D_x\varphi_\tau(x^*, t_0) = \Phi_\tau(x^*, t_0)$$

for which the characteristic multipliers of a non-autonomous system are equal to the eigenvalues of the matrix $\Phi_\tau(x^*, t_0)$.

The Lyapunov exponents are a generalization of the eigenvalues and of the characteristic multipliers; they allow the stability of an equilibrium point to be analyzed as well as that of periodic, quasi-periodic, or chaotic solution.

For an autonomous time-continuous system, after considering the initial condition $x_0 \in R^n$, let $m_1(t),....,m_n(t)$ be the eigenvalues of $\Phi_t(x_0)$; the Lyapunov exponents of x_0 are given by the relation:

$$\lambda_i := \lim_{t\to\infty} \frac{1}{t} \ln\left|m_i(t)\right|, \qquad i = 1,...n \tag{8.29}$$

The coefficients λ_i are generally considered in non ascending order $\lambda_1 \geq \lambda_2 ... \geq \lambda_n$.

From relation (8.24), it is seen that the Lyapunov exponents depend on the initial condition x_0; however, since these are referred to the regime behavior of the system, every point belonging to the basin of attraction of an attractor will have the same Lyapunov exponent; they are therefore best referred to as *Lyapunov exponents of an attractor.*

A first condition regarding Lyapunov exponents is obtained from the system dissipation hypothesis; in fact, in this type of system, the flow must be contractive and therefore the sum of all the Lyapunov exponents of an attractor must be negative:

$$\sum_{i=1}^{n} \lambda_i < 0 \tag{8.30}$$

The positive value of a Lyapunov exponent indicates the presence of an expansion phenomenon in the system flow; thus, in view of what was said above, this condition univocally identifies a non-linear dynamic system as being chaotic; furthermore, it has been proved [17] that for every limit set other than an equilibrium state, at least one Lyapunov exponent is always zero.

Since the sum of all the Lyapunov exponents must be negative, and at least one of these must be zero, it follows that for a dynamic system to be chaotic, it must be at least a third order one.

In Table 8.1, the various attractors of non-linear dynamic systems are classified according to the Lyapunov exponents.

Table 8.1. Classification of attractors by means of Lyapunov exponents

Regime solution	**Limit Set**	**Poincaré map**	**Lyapunov exponents**	**Dimensions**
Equilibrium state	Point		$0 \geq \lambda_1 \geq ... \geq \lambda_n$	0
K-periodic solution	Loop	Fractal set	$\lambda_1 = 0$ $0 \geq \lambda_2 \geq ... \geq \lambda_n$	1
Chaotic behavior	Fractal set	One or more points	$\lambda_1 > 0$ $\sum\lambda_i < 0$	Non integer

8.5 Dimension of the Attractors

The concept of *dimensions* is widely known in vastly different fields: the dimension of a Euclidean space is the minimum number of coordinates needed for univocally specifying a point in space; the dimension of a dynamic system is equal to the minimum number of variables needed for describing completely the state of the system; the dimension of a manifold is equal to the dimension of the Euclidean space in which the manifold can be locally enclosed [1]. However, none of these definitions can be applied to determining the dimension of a strange attractor. Generically, a set having an non integer dimension is said to be a fractal set: almost all strange attractors are fractal.

In the previous paragraphs, the concept of *differomorphism* was introduced; now we will introduce another concept of differential geometry [18] which will be used below for defining the dimension of a strange attractor: *the manifold*.

Definition 8.16. *A k-dimensional manifold M is a set of points which can be locally assimilated to a subset of R^k. More formally, M is a k-dimensional manifold if for every point $x \in M$ of this it is possible to determine an open neighborhood U(x) which is a differomorphic copy of some open R^k.*

For example, a circle is a mono-dimensional manifold, a sphere is a two-dimensional manifold; a square is not a manifold in that it is not continuous, with continuous derivatives at the corners.

Many different definitions of dimension can be found in the literature, and they can be used for chaotic system attractors [21-22]; below, we will discuss some of the most commonly used ones.

The most simple of the various definitions of dimension is the capacity dimension d_{cap}. Let the attractor be covered with elements of equal volume such as, for example, cubes and spheres, and let ε be the elementary dimension of such elements, the side for a cube or the radius for a sphere. Then let $N(\varepsilon)$ be the minimum number of elements needed for covering the attractor.

If the attractor is a d-dimensional manifold with a whole d, the number of elements needed for covering the attractor is inversely proportional to ε^d [1], and hence, given k the proportionality constant, we obtain:

$$N(\varepsilon)=k\varepsilon^{-d} \tag{8.31}$$

The capacity dimension d_{cap} is obtained by solving (8.31) with respect to d and considering the limit for ε to be tending to zero:

$$d_{cap}=\lim_{\varepsilon\to 0}\frac{\ln N(\varepsilon)}{\ln\left(\frac{1}{\varepsilon}\right)} \tag{8.32}$$

The *capacity dimension* is defined only if the limit (8.32) is finite.

Since a manifold is locally similar to a Euclidean space, its dimension will be equal to the topological one of the space; if the attractor is not a manifold, the value of d_{cap} will not be a non integer one.

Example 8.1. Unitary Interval. Let us consider a segment of unitary length and let us determine its dimension, covering it with elements of length $\varepsilon = (1/3)^k$. We therefore need $N(\varepsilon) = 3^k$ elements in order to cover completely the interval [0,1]. From relation (8.32) we obtain $d_{cap} = 1$, and thus the segment with unitary length has a dimension of 1.

Example 8.2. Cantor set. A Cantor set can be constructed iteratively starting from the unitary interval, dividing each continuous portion of it into three equal parts and eliminating the central part [1], as shown in Figure 8.7.

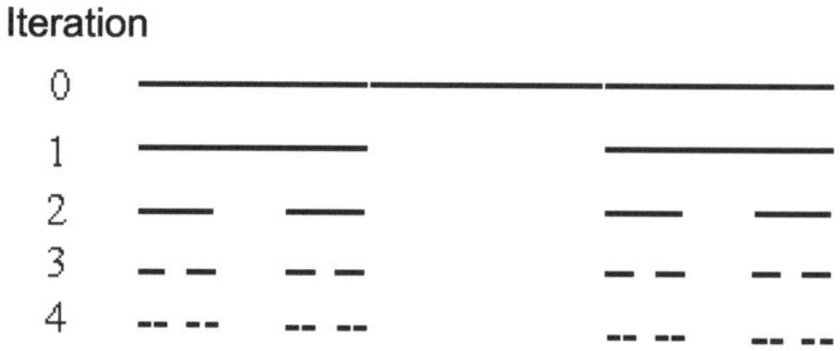

Iteration
0
1
2
3
4

Figure 8.7. Construction of the so-called middle-third Cantor set

Covering the set again with segments of length $\varepsilon = (1/3)^k$, we obtain $N(\varepsilon) = (2/3)^k$, and thus:

The *capacity dimension* is a function exclusively of the metric properties of the attractor and does not take into account the temporal evolution of the trajectories. The information dimension d_I is defined in function of the relative frequency with which a typical trajectory visits the region containing the attractor.

Covering the attractor, as in the case of the capacity dimension, with $N(\varepsilon)$ elements of volume, the *information dimension* is given by the following relation:

$$d_I = \lim_{\varepsilon \to 0} \frac{H(\varepsilon)}{\ln(1/\varepsilon)} \tag{8.33}$$

where the term $H(\varepsilon)$ represents the trajectory entropy [1], that is to say the quantity of information required

$$H(\varepsilon) = -\sum_{i=1}^{N(\varepsilon)} P_i \ln P_i \tag{8.34}$$

A further definition of dimension of an attractor, based on its statistical properties, is the correlation dimension d_c. Again with respect to the $N(\varepsilon)$ covering of the attractor, we have:

$$d_C = \lim_{\varepsilon \to 0} \frac{\ln \sum_{i=1}^{N(\varepsilon)} P_i^2}{\ln \varepsilon} \tag{8.35}$$

where the term P_i again represents the statistical frequency with which the i-th element of volume is visited by the typical trajectory contained in the attractor. It can be proved that among the definitions of dimension mentioned here, the following relation holds [1]:

$$d_c < d_I < d_{cap}$$

The dimension can be used for classifying the attractors and in particular for identifying the strange attractors; in fact, the latter possess a non-whole dimension.

However, this classification is of no great practical utility because of the difficulties inherent in calculating the dimensions; in fact, it often proves difficult to distinguish a whole value from a non-whole one, due to numerical problems of the computing algorithms [19].

The main use of dimensions consists, instead, in determining the minimum number of variables needed for identifying the attractor: therefore, they afford a measurement of the complexity of an attractor.

Yet another definition of the dimension of an attractor is based on a knowledge of the Lyapunov exponents of the attractor and was devised by Kaplan [20]; this is known as the Lyapunov dimension d_L.

Let us consider the set of Lyapunov exponents of an attractor $\lambda_1 \geq ... \geq \lambda_n$, and let j be the greatest whole for which $\lambda_1 + ... \lambda_j \geq 0$; the *Lyapunov dimension* d_L is defined as:

$$d_L = j + \frac{\lambda_1 + ... + \lambda_j}{|\lambda_{j+1}|} \tag{8.36}$$

If it is not possible to determine a value of j for which (8.36) holds, d_L is defined as being equal to zero: this is the case with an equilibrium point.

For dissipative systems, the summation of the Lyapunov exponents is negative, which guarantees that j will always be less than n.

By way of example, for a limit cycle we will have:

$$\lambda_1 = 0 \geq \lambda_{12} \geq ... \geq \lambda_n$$

and hence from (8.36) it follows that the dimension of this attractor is equal to one.

8.6 Circuit Models of Chaotic Dynamic Systems

This section will give examples of circuit diagrams and the values of the components to be used for realizing the electric models of some previously mentioned dynamic systems.

The components used were: *μA741*- and *TL082*-type operational amplifiers and *AD633JN* analog multipliers.

8.6.1 The Duffing Oscillator

The *Duffing oscillator*, the model of which is given from (8.13), was realized according to the circuit diagram reported in Figure 8.8.

The circuit uses two multipliers to realize the function x^3. The component values of this circuit are listed in Table 8.2.

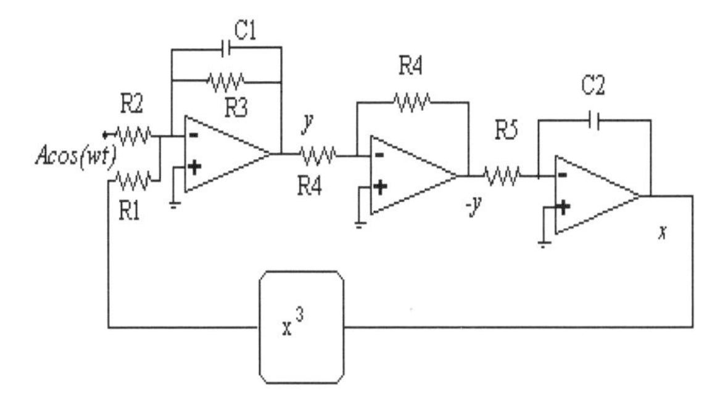

Figure 8.8. Block diagram of the Duffing oscillator electric circuit

Table 8.2. List of components of the Duffing oscillator

The Duffing oscillator: list of components	
Resistors	
R1=200 Ω	R2=2.6 KΩ
R3=400 KΩ	R4=2.2
R5=20 KΩ	
Capacity	
C1=C2=100 μF	

8.6.2 Chua's Circuit

This circuit is composed of a linear section including the elements L, C_1, C_2, and R, and a non-linear one represented by Chua's diode N_R. The circuit diagram and the features of the non-linear element are reported in Figures 8.9 and 8.10 respectively.

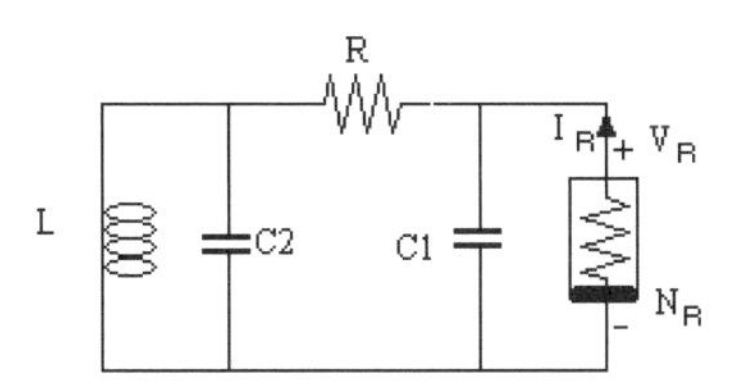

Figure 8.9. Chua's circuit

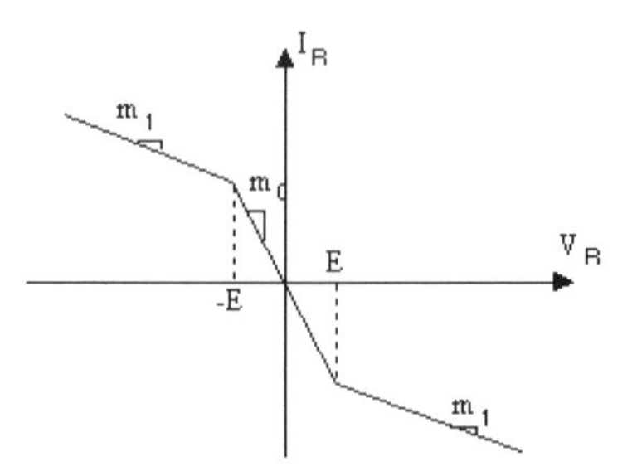

Figure 8.10. Chua's diode characteristic

The circuit of the non-linear component was realized using the saturation characteristics of two inverting amplifiers with different gain, connected in parallel according to the diagram of Figure 8.11.

To realize the L=18mH inductor a negative impedance converter (NIC) was used, the diagram of which is shown in Figure 8.12, as an inductance simulator, thereby reducing the overall dimensions of the circuit.

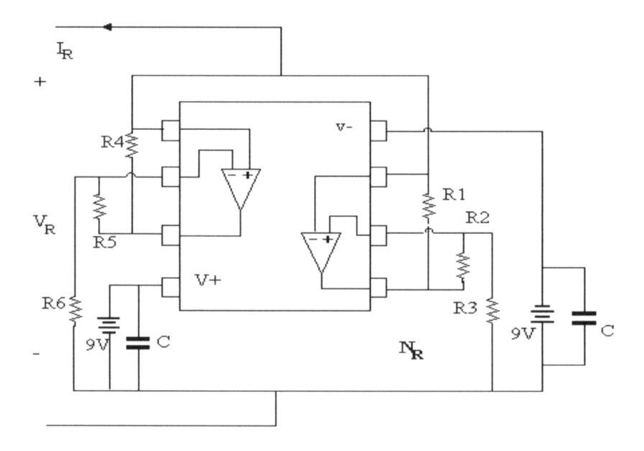

Figure 8.11. Circuit diagram of Chua's diode

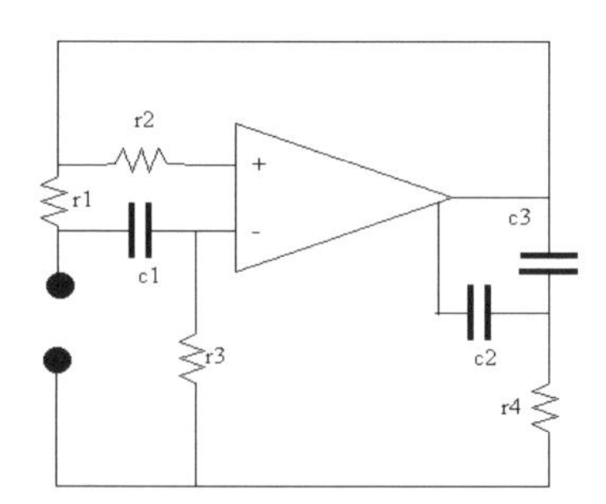

Figure 8.12. Circuit diagram of a NIC

The values of the components used are reported in Table 8.3.

Table 8.3. Chua's circuit components values

Chua's circuit: components values	
Resistors	
R1=R2=220 Ω	R3=2.2 KΩ
R4=R5=22 KΩ	R6=3.3 KΩ
r1=18 Ω, r2=r3=100 Ω	r4=10 Ω
Variable resistor	
R=2 KΩ	
Capacity	
C1=10 *n*F	C2=100 *n*F
C1=0.1μF	c2=33*p*F, c3=330*p*F

8.6.3 The Lorenz Circuit

The circuit diagram of the *Lorenz system* electric model described by Equations 8.15 is shown in Figure 8.13. The values of the components used are listed in Table 8.4.

Table 8.4. List of components of the Lorenz circuit

The Lorenz circuit: components values
Resistors
R1=R2=R3=R4=R5=R6=R8=R10=R11=R12=R16=R17 =10 KΩ R7=3.3 KΩ R9=R13=R14=R15=R19=1 KΩ R18=39 KΩ
Capacity
C1=C2=C3=100 *n*F

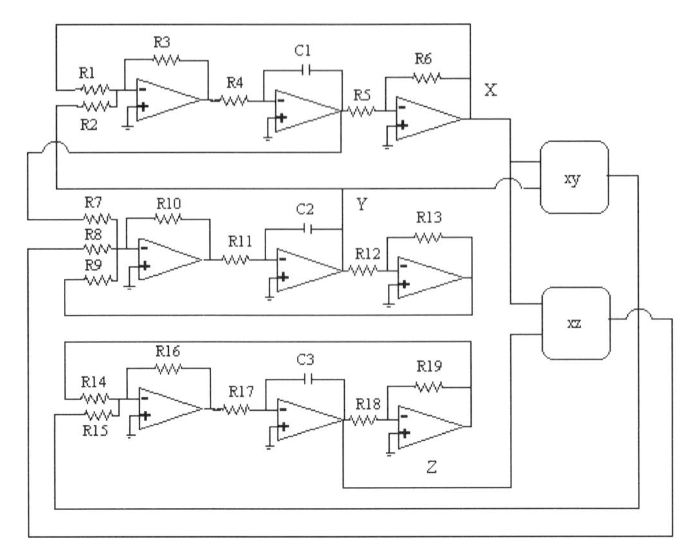

Figure 8.13. Diagram of the Lorenz system circuit

8.7 Generating Chaotic Phenomena by CNN

This section describes an application of CNNs to chaotic systems. In particular, it will be shown how *chaotic behaviors* can be generated by CNNs.

Let us consider particular autonomous CNN configurations able to generate chaotic-type dynamic behaviors of different kinds [23]. These configurations require a modification of the individual cell state equations (7.8) by introducing *state interaction terms* between adjacent cells. The corresponding architecture is named a *state-controlled neural network* (SC-CNN).

The interaction between the cells occurs through the outputs and inputs of the adjacent cells (weighted by means of the feedback and control templates).

The SC-CNN equations are:

$$\begin{aligned} C\dot{x}_{ij}(t) &= -\frac{1}{R}x_{ij}(t) + \sum_{C(K,l)\in N_r(i,j)} A(i,j,k,l)y_{Kl} \\ &+ \sum_{C(K,l)\in N_r(i,j)} B(i,j,k,l)u_{Kl} + \sum_{C(K,l)\in N_r(i,j)} C(i,j,k,l)x_{Kl} + I \\ x_{ij}(0) &= x_{ij0} C > 0, R_x > 0, 1 \le i \le M, 1 \le j \le N \end{aligned} \tag{8.37}$$

The constants $C(i,j,k,l)$ are named state templates.

It is readily proved [24-25] that starting from the state equations of some of the best known chaotic circuits (e.g., Chua's circuit, the Colpitts oscillator, the Nishio-Ushida circuit, Saito's circuit, *etc.*), it is enough to perform a mapping operation of the state equations of such circuits in the form of (7.8) in order to obtain the corresponding templates $\{A(i,j;k,l)\}$, $\{S(i,j;\ k,l)\}$ and $\{I_{ij}\}$ of the cellular neural network.

The basic idea justifying the use of SC-CNNs for generating chaotic dynamics is their possibility of being easily and rapidly programmed by means of the template parameters transconductance [26] for generating chaotic signals of various kinds.

Incidentally, another field of research undergoing full development is the realization of analog cryptographic systems based on the exploitation of chaotic oscillators, which are employed for realizing the encoders and decoders [27-28].

In this framework, a potential listener will be able to reconstruct a message codified in this way if and only if he knows exactly the type of dynamics employed for the masking. That means that to every chaotic dynamic there will correspond a possible coding of the message that one wishes to mask. It is thus clear that the use of a programmable chaotic dynamics generator such as the state-controlled CNN affords evident advantages when realizing a cryptographic system which makes use of encoders and decoders [27].

8.7.1 Implementing a Chua Circuit by Means of CNN

An applicational example of the strategy proposed will be given below with respect to the previously examined Chua circuit.

Chua's circuit [29] is well known from the literature as being one of the first circuits able to generate various chaotic attractors. Its dimensionless circuit equations are:

$$\begin{cases} x = \alpha(y - h(x)) \\ y = x - y + z \\ z = -\beta y - \gamma \end{cases} \tag{8.38}$$

$$h(x) = m_1 x + 0.5(m_0 - m_1)(|x+1| - |x-1)|$$

where x,y, and z represent the state variables (connected to the tensions on the two condensers and to the inductor current, Figure 8.10), whereas α, β, γ, m_0, and m_1 are the parameters (connected to the circuit's component values).

Various chaotic attractors can be obtained by varying the Chua's circuit parameters.

Now, let us consider a SC-CNN composed of three cells only. Given the low number of cells, the notation can be simplified by considering them arranged, for example, in one line only, and thus employing one index only rather than pairs of indices. The input contribution should be ignored (control templates zero). Equations (8.37) will now become:

$$C\dot{x}_i(t) = -\frac{1}{R}x_i(t) + \sum_{C(K)\in N_r(i)} A(i,k)y_K + \sum_{C(K)\in N_r(i)} C(i,k)x_K + I \tag{8.39}$$

$$i \le 1 \le 3$$

For the sake of simplicity, we assume that R_x and C are equal to 1. Then if we want x_1, x_2, and x_3 to coincide with x, y, and z, respectively, of the Chua oscillator, the following relations must hold:

$$\begin{aligned}
&A(1;2)=A(1;3)=A(2;2)=A(2;3)=A(3;2)=A(3;3)=A(2;1)=A(3;1)=0;\\
&A(1;1)=\alpha(m_0-m_1);\\
&C(1;1)=1-\alpha m_1;\ C(1;2)=\alpha;\ C(2;1)=C(2;3)=1;\ C(3;2)=-\beta;\\
&C(3;3)=1-\gamma;\ C(1;3)=C(3;1)=C(2;2)=0;\quad I1=I2=I3=0;
\end{aligned} \tag{8.40}$$

When the SC-CNN templates (8.39) are (8.40), then they will be equal to the Chua oscillator in the sense that both will have precisely the same dynamics.

However, some advantages of realizing this kind of dynamics based on the SC-CNN should be emphasized. First of all, a realization based on a SC-CNN does not contain any inductors but only condensers. In addition, the three condensers are all equal. These two points are of considerable importance when one thinks of an integrated realization.

In fact, it is well known that the realization of inductors is prohibitive; what is more, whereas the nominal values of the inductors are subject to relatively high realization uncertainties (sometimes as much as 10%), on the other hand it is easy to realize condensers whose capacity is in a much better defined relationship (uncertainty less than 1%) [26].

Finally, it can be observed that a great variety of attractors can be realized by varying α, β, γ, m_0, and m_1; in the classical realization of Figure 8.14, this would require negative values for the components (negative inductance, negative capacity, *etc.*).

For the circuit, that would involve complications. Instead, in a SC-CNN, since the template values correspond to the piloted generator gains, they can assume both positive and negative values without thereby giving rise to any problems.

One possible experimental implementation [30] of SC-CNNs 8.39 and 8.40 carried out with discrete components (where only 4 common operative amplifiers are needed) is given in Figure 8.14.

The component values were chosen for a double scroll realization.

The non-linear term required for realizing y_1 was obtained by designing an amplifier stage that includes A2 so that its output is saturated when its input x_1 has an absolute value greater than 1. The templates correspond to the gains of the remaining adder/subtracter amplifiers [30].

It should be mentioned that a patent has been taken out for realizing Chua's oscillator by means of SC-CNNs.

We can proceed in a similar way to generate the chaotic dynamics of various systems [31].

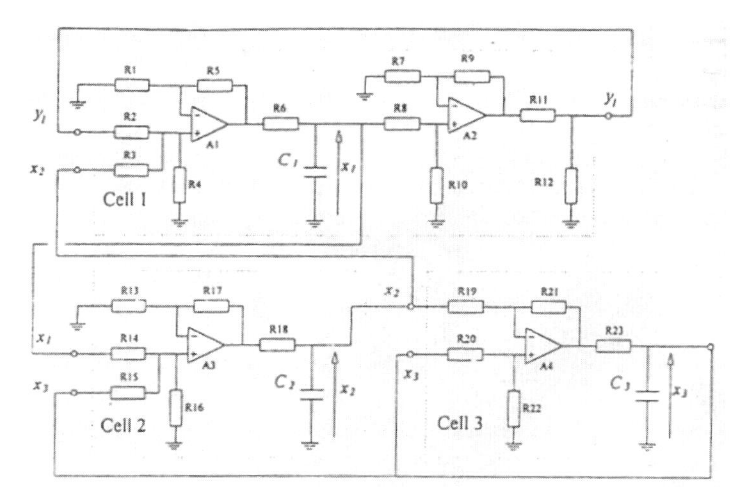

Figure 8.14. Implementation of the SC-CNN cells for realizing the double scroll

Components:

Cell 1: R1=4K; R2=13.2K;R3=5.7K; R4=20K; R5=20K; R6=1K; R7=75K; R8=75K; R9= 1M; R10=1M; R11=12.1K; R12=1K; C1=100n;
Cell 2: R13=51.1K; R14=100K; R15=100K; R16=100K; R17=100K; R18=1K; C2=100n;
Cell 3: R19=8.2K; R20=100K; R21=100K; R22=7.8K; R23=1K; C3=100n;
Feed : V_{cc}=+15V; V_{ee}=-15V.

The attractor obtained when realizing a SC-CNN determined in this way is reported in Figures 8.15.

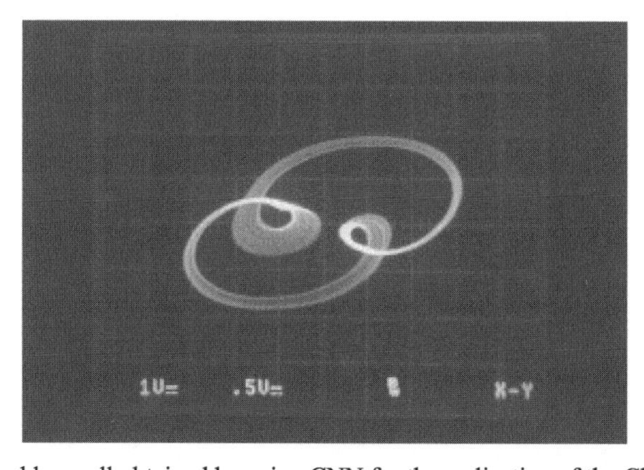

Figure 8.15. Double scroll obtained by using CNN for the realization of the Chua circuit

8.7.2 SC-CNNs as a Platform for Generating Other Chaotic Dynamics

We saw in the previous section how, by appropriate choice of templates, a circuit made up of three SC-CNN cells can generate Chua's oscillator dynamics. Can other dynamics be generated with the same approach? The answer is definitely affirmative. In fact, it has been proved both analytically and experimentally [32] that with the same approach a great number of other non-linear dynamics can be generated. In this section, we will list some of the most significant cases of all those considered. The reader should consult the bibliographical references for theoretical details [32].

Recently [33], it has been shown how a common three-pointed_oscillator, erroneously defined as a Colpitts one on account of the non-linearity due to the active device (BJT), may display anomalous behavior. Although it has not been proved in theory that this anomalous behavior is really chaotic, the existence of a strange attractor has been shown both in simulation and in experimental conditions [33].

The attractor obtained from a SPICE oscillator simulation is reported in Figure 8.16. A sufficiently accurate model for describing such a circuit is the following:

$$\begin{cases} C\dfrac{dV_{CE}}{d\tau} = I_L - I_C \\ C\dfrac{dV_{BE}}{d\tau} = -\dfrac{V_{EE}+V_{BE}}{R_{EE}} - I_L - I_B \\ L\dfrac{dI_L}{d\tau} = V_{CC} - V_{CE} + V_{BE} - I_L R_L \end{cases}$$

$$I_B = \begin{cases} 0 & se \quad V_{BE} \le V_{TH} \\ \dfrac{V_{BE}-V_{TH}}{R_{ON}} & se \quad V_{BE} > V_{TH} \end{cases}$$

$$I_C = \beta_F I_B \tag{8.41}$$

where V_{CE}, V_{BE} e I_L are the state variables. The anomalous functioning is observed when: $V_{TH}=0.75V$, $R_{ON}=200\Omega$, $R_L=35\Omega$, $L=98.5\mu H$, $C=54nF$, $R_{EE}=400\Omega$, $\beta_F=256$, $V_{EE}= -5V$, $V_{CC}= 5V$.

In this case, too, as in that of the Chua oscillator, this dynamic functioning can be reproduced in a SC-CNN formed by three cells. This time, the templates will be [32]:

$$C(1;1)=1;\ C(1;2)=-\frac{\beta_F K R_L}{aR_{ON}V_{CC}};C(1;3)=1;\ A(1;2)=\frac{\beta_F K R_L}{aR_{ON}V_{CC}};$$

$$C(2;2)=1-\frac{R_L}{R_{EE}}-\frac{R_L}{R_{ON}};\ C(2;3)=-\frac{aV_{CC}}{K};A(2;2)=\frac{R_L}{R_{ON}};C(3;1)=-\frac{R_L^2C}{L};$$

$$C(3;2)=\frac{KR_L^2C}{aLV_{CC}};\ C(3;3)=1-\frac{R_L^2C}{L};\ I_2=\frac{R_L(b-aV_{EE})}{KR_{EE}};\ I_3=\frac{R_L^2C}{L}-\frac{bR_L^2C}{aLV_{CC}};$$

$$a = b = 4; \; K = aV_{TH} + b; \tag{8.42}$$

all the other unlisted coefficients being zero.

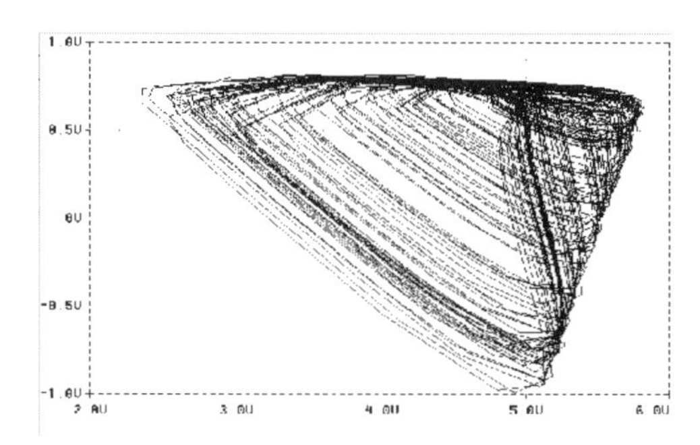

Figure 8.16. Three-pointed oscillator in a SPICE simulation (the state variables V_{CE} and V_{BE} are reported in the x- and y-coordinates, respectively)

The attractor observed experimentally in the realization with a SC-CNN is shown in the photograph of Figure 8.17.

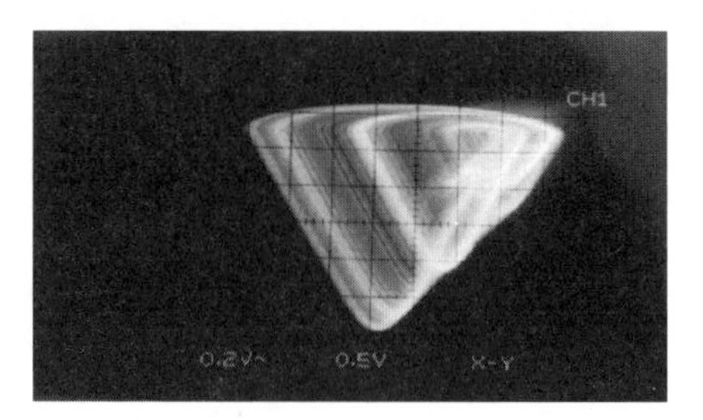

Figure 8.17. The attractor observed in the realization with a SC-CNN (x_1 and x_2 are reported in the x- and y-coordinates, respectively)

Another example is the *hyperchaotic* Saito circuit. It is well known that a circuit is chaotic when it has a positive Lyapunov exponent, whereas it becomes *hyperchaos* if the positive Lyapunov exponents are more than one [29].

It is equally known that a necessary condition for hyperchaos to occur is that the circuit or system be of an order greater than three [29].

The circuit proposed by Saito in [34] belongs to this category; it is a fourth-order circuit and displays chaotic and hyperchaotic behaviors.

The equations of this circuit can be expressed in the dimensionless form:

$$\begin{cases} \dot{x} = -z - w \\ \dot{y} = \gamma(2\delta y + z) \\ \dot{z} = \rho(x - y) \\ \varepsilon\dot{w} = z - h(w) \end{cases} \tag{8.43}$$
$$h(w) = w - (|w+1| - |w-1|)$$

where x, y, z, and w are the state variables and γ, ρ, δ, and ε are the parameters.

The circuit displays two Lyapunov exponents when $\gamma = 1$, $\rho = 14$, $\delta = 1$, and $\varepsilon \to 0$ (in practice it is sufficient that $\varepsilon = 10^{-2}$). In this case, four SC-CNN cells will be needed (one for every equation/state variable). The following templates will be required:

$$C(1;1) = 1;\ C(1;3) = C(1;4) = -1;\ C(2;2;) = 1 + 2\gamma\delta;\ C(2;3) = \gamma;$$
$$C(3;1) = \rho;\ C(3;2) = -\rho;\ C(3;3) = 1;\ A(4;4) = 2/\varepsilon;\ C(4;1) = 1/\varepsilon;$$
$$C(4;4) = 1 - 1/\varepsilon; \tag{8.44}$$

with all the remaining coefficients being zero.

The hyperchaotic attractor observed experimentally with a 4-cell SC-CNN with the (12) templates calculated for $\gamma = 1$, $\rho = 14$, $\delta = 1$, and $\varepsilon = 10^{-2}$ is shown in the photograph of Figure 8.18.

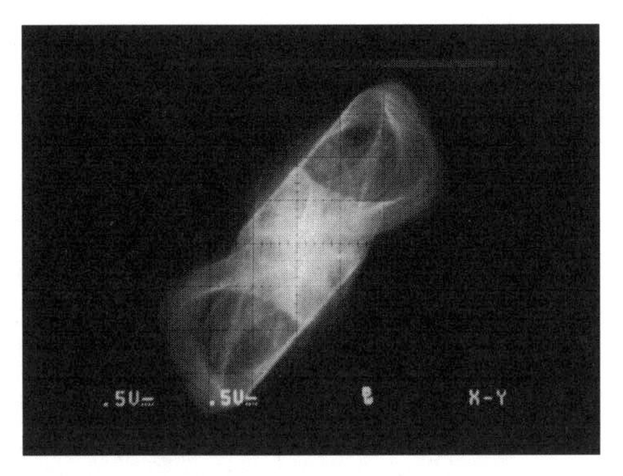

Figure 8.18. The hyperchaotic Saito circuit attractor observed in a realization with a SC-CNN (x_1 and x_2 are reported in the x- and y-coordinates, respectively)

In [35], a new family of attractors was presented, the so-called *double scrolls*, which include, as a particular case, the well-known double scroll of Chua's oscillator.

Generally, n-double scroll resembles a set of n different double scrolls of various dimensions, one inside the other, and all centered in the origin.

In order to obtain these attractors the non-linearity of Chua's diode must be modified.

The non-linearity proposed in [35] is rather difficult to realize in a circuit, and thus in [25] a different non-linearity was introduced with a view to realizing attractors of the same kind.

It is the $h(x)$ bold solid line linear function shown in Figure 8.19 [25].

Figure 8.20 shows, by way of example, the 2-double scroll and the 3-double scroll obtained in simulation with the non-linearity proposed in [25].

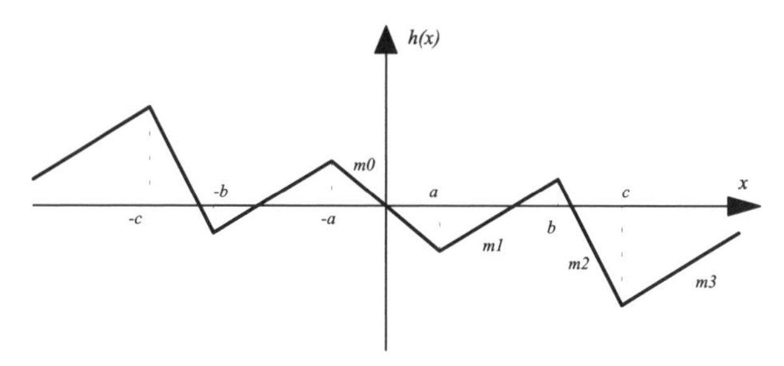

Figure 8.19. Non-linearity of Chua's diode for obtaining a 2-double scroll

Finally, for the systems obtained in this way, a realization based on SC-CNNs has been shown.

With this in view, it was necessary to adopt a different output non-linearity for the SC-CNN cells; the new non-linearity in the case of a 2-double scroll is reported in Figure 8.21a.

An experimentally realized 2-double scroll with a SC-CNN is shown in the photograph of Figure 8.21b.

Theoretical and implementational details can be found in [25].

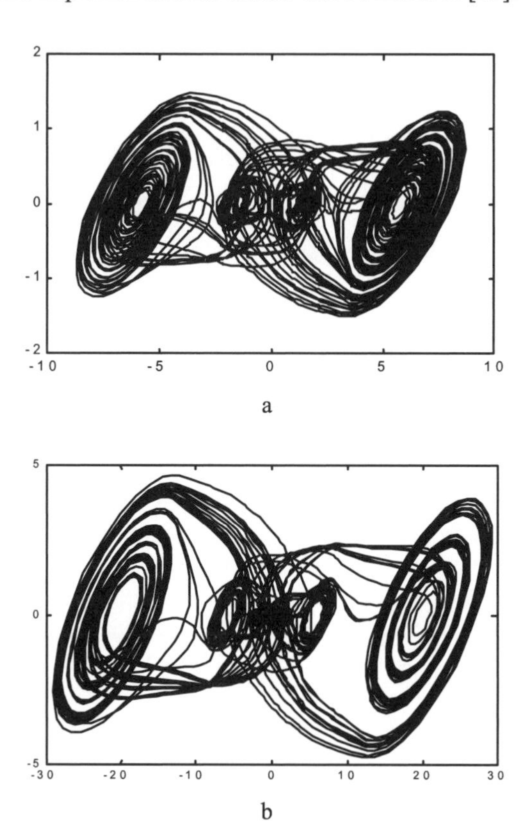

Figure 8.20. a. 2-double scroll; b. 3-double scroll

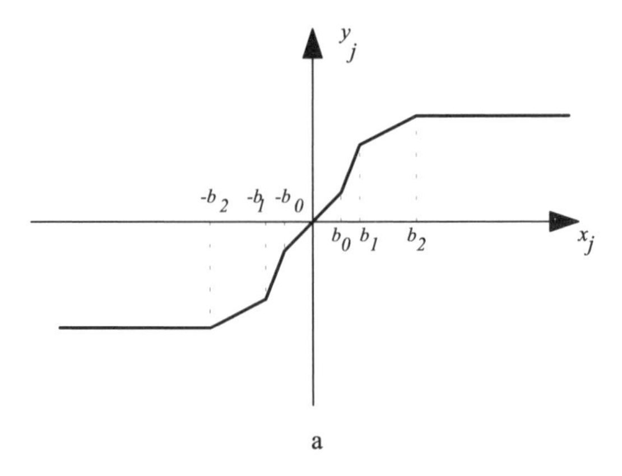

a

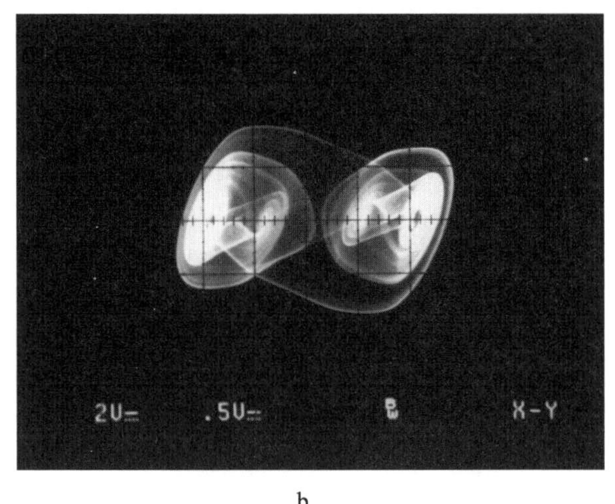

b

Figure 8.21. a. output non-linearity; b. 2-double scroll obtained by means of an experimental realization with a SC-CNN (x_1 and x_2 are reported in the x- and y-coordinates, respectively)

The number of attractors that can be obtained with circuits based on a SC-CNN is extremely vast. For the sake of brevity, we have therefore preferred to report only those described previously, purely by way of demonstration, while the reader should consult the literature quoted in the bibliography [23].

In conclusion, we would like to emphasize that the use of a SC-CNN cannot be limited only to reproducing the dynamics of other circuits. The dynamic behaviors outlined here are merely an example of the immense amount of functioning possibilities that can be obtained; many other dynamics not observed elsewhere have been obtained experimentally and just as many may be revealed in the future. Thus, the SC-CNN is a new paradigm for generating non-linear dynamics.

This potentiality, together with the possibility of implementing programmable SC-CNNs, makes it evident that there is the chance of producing, in a simple way, efficiently, and at low cost, generators of non-linear oscillations that can be programmed and employed, for example, in analog cryptography [27] or in spread spectrum communications [27-28].

8.8 References

1. T S Parker, Chua LO. Practical Numerical Algorithms for Chaotic Systems. Springer-Verlag, 1989
2. Eckmann JP, Ruelle D. Ergodic Theory of Chaos and Strange Attractors. Reviews of Modern Physics Part I. 1985; 57: 3
3. Thompson JMT, Stewart HB. Nonlinear Dynamics and Chaos. John Wiley and Sons, 1989
4. Lorenz EN. Deterministic Nonperiodic Flow. Journal Atmospheric Science. 1963; 20; 130
5. Rasband SN. Chaotic Dynamics of Nonlinear Systems. John Wiley & Sons, 1990
6. Hasler MJ. Electrical Circuits with Chaotic Behavior. l Proceedings of the IEEE, Aug. 1987
7. Ikeda K. Chaotic Itinerancy. International Symposia on Information Sciences, IIzuka, July 1992
8. Gleick J. Chaos: Making a New Science. Viking Press, 1987
9. Mayer-Kress G, .Choi I, Weber N, Bargar R, Hübler A. Music Signals from Chua's Circuit. IEEE Transactions on Circuits and Systems II. 1993; 40
10. Heileman GL, Abdallah C, Hush D, Baglio S. Chaotic Probe Strategies in Open Addressing Hashing. 1993 International Symposium on Nonlinear Theory and its Applications, Hawaii, Dec, 1993
11. Chaos Simulator and Application to Washing Machine, GOLDSTAR, Tech. Rep., July 1993
12. Dedieu H, Kennedy MP, Hasler M. Chaos Shift Keying: Modulation and Demodulation of a Chaotic Character Using Self-Synchronizing Chua's Circuits. IEEE Transactions on Circuits and Systems II. 1993; 40
13. Barnsley MF. Fractals Everywhere. Academic Press Professional, 1993
14. Guckenheimer J, Holmes P. Nonlinear Oscillation, Dynamical Systems and Bifurcations of Vector Fields. Springer-Verlag, 1983
15. Fornasini E, Marchesini G. Appunti di Teoria dei Sistemi. Edizioni Libreria Progetto, Padova, 1985
16. Oseledec VI. A Multiplicative Ergodic Theorem. Lyapunov Characteristic Numbers for Dynamical Systems. Trans. Moscow Math. Soc. 1968; 19: 97
17. Haken H. At least One Lyapunov Exponent Vanishes if the Trajectory of an Attractor does not Contain a Fixed Point. Physics Letter. 1983; 94A(2): 71-72
18. Guillemin V, Pollack A. Differential Topology. Prentice-Hall, Englewood Cliffs 1974
19. .Grassberger P, Procaccia I. Dimension and Entropies of Strange Attractors from a Fluctuating Dynamics Approach. Physica D. 1984; 13
20. Kaplan JL, Yorke JA. Chaotic Behavior of Multidimensional Difference Equations. Lecture Notes in Mathematics Springer-Verlag, 1979; 228-237
21. Badii R, Politi A. Statistical Description of Chaotic Attractors: The Dimension Function. Journal of Statistical Physics. 1985; 40 (5/6): 725-750

22. Mayer-Kress G. (editor). Dimension and Entropies in Chaotic Systems. Springer-Verlag, 1986
23. Arena P. Baglio S, Fortuna L, Manganaro G. State Controlled CNN: A New Strategy for Generating High Complex Dynamics. IEICE trans. On Fundamentals October 1996; E79-A: 10
24. Arena P, Baglio S, Fortuna L, Manganaro G. Chua's Ciurcuit can be Generated by CNN Cells. IEEE Trans. On Circuits and Systems-Part I. February 1995; 42: 2
25. Arena P. Baglio S, Fortuna L, Manganaro G. Generation of N-Double Scroll Via Cellular Neural Networks. Int. Journal of Circuit Theory and Application 1966; 24
26. Geiger RL, Allen PE, Strader NR. VLSI: Design Techniques for Analog and Digital Circuits. McGraw-Hill, 1990
27. Arena P, Baglio S, Fortuna L, Manganaro G. Experimental Signal Transmission Using Synchronised State Controlled Cellular Neural Networks. IEE Electronics Letters. 1996; 32: 362-363
28. Caponetto R, Lavorgna M, Occhipinti L. Cellular Neural Networks in Secure Transmission Applications. 4th IEEE Int. Workshop on Cellular Neural Networks and their Appl.s, Seville, 1996; 411-416
29. Chua LO (editor). Special Issue on Chaotic System, Proc. of IEEE. August 1987
30. Arena P, Baglio S, Fortuna L, Manganaro G. A Simplified Scheme for the Realisation of the Chua's Oscillator by Using SC-CNN Cells. IEE Electronics Letters. 1995; 31: 1794-1795
31. Roska T, Wu CW, Balsi M, Chua LO. Stability and Dynamics of Delay-Type General and Cellular Neural Networks. IEEE Transactions on Circuits and Systems-I: Fundamental Theory and Applications, June 1992; 39: 6
32. Arena P, Baglio S, Fortuna L, Manganaro G. How State Controlled CNN cells generate the dynamics of the Colpitts-like Oscillator. IEEE Trans. on Circuits and Systems - part I, July 1996; 43
33. Kenney MP. Chaos in the Colpitts Oscillator. IEEE Trans.on Circuits and Systems - part I, 1994; 41: 771-774
34. Saito T. An approach toward higher dimensional hysteresis chaos generators. IEEE Trans. on Circuits and Systems 1990; 37: 399-409
35. Suykens JAK, Vandewalle J. Generation of n-Double Scrolls (n=1,2,3,4,..). IEEE Trans. on Circuits and Systems - part I, 1993; 40: 861-867

9. Neuro-fuzzy Networks

9.1 Introduction

One of the most important research themes, in the sense of intelligent processing techniques hybridization, is the *neuro-fuzzy* approach. The birth of this kind of system is mostly connected with the attempt to unify the advantages of neural and fuzzy techniques using one *hybrid architecture* only, often referred to as *fuzzy neural networks* (FNN).

Of the fields of major interest for applying such technologies there are the automatic control of dynamic systems and the identification of static and dynamic systems.

The main idea consists in implementing a system of fuzzy rules in a neural-type architecture. In fact, in this way it is possible to train the neuro-fuzzy network with a suitable learning algorithm which, as for neural networks, iteratively adapts its own parameters to optimizing a cost index. FNNs can thus be used, starting from a set of experimental input-output data, for identifying a non-linear function. At the end of the training, the value of the interconnection weights can be easily traced back to the value of the fuzzy set parameters of the antecedents and the consequents of the fuzzy rules system. A FNN can be used both as a neural model and as a fuzzy rules system which can be transferred to a generic fuzzy chip.

Below, we will describe the basic characteristics of FNNs with respect to neural networks and fuzzy systems:

- FNNs realize an *automatic procedure* for obtaining in the same time both the consequents and antecedents of a set of fuzzy rules starting from a system's set of input-output data; in addition, they allow us to appropriately modify the shape of the membership functions as well.
- FNNs require a *small number of parameters* with respect to the number of connections in a MLP; moreover, the number of neurons in a FNN is wholly determined by the number of membership functions chosen for each input variable.
- FNNs require a *short training phase* thanks to the fact that a meaningful initial configuration for the parameters can be established.
- FNNs allow us to *incorporate the knowledge* of an expert regarding the choice of topology.

- FNNs lead us to determine a system model which is *easily comprehensible*, unlike the model obtained with a MLP; in fact, neural networks reach their own limits precisely because the knowledge acquired by a neural network consists in a set of interconnection weights which is not simply interpretable by a human operator. Instead, a fuzzy rules system is always *transparent* in the sense that a human operator can easily *read* the knowledge base of the fuzzy system and interpret its behavior when faced by a given situation.

9.2 Neuro-fuzzy Network Architecture

Several different types of FNN structure have been proposed in the literature [1-2]; these differ from each other with regard to the shape of the consequent part of the fuzzy rules, the type of inference rules, and the shape of the membership functions. One possible *FNN structure*, described in [1], [2], [3], implements a set of IF-THEN rules with a sigmoidal-type membership function and crisp consequents. To obtain a learning algorithm for a FNN with the gradient technique, the inference rule must lead to a crisp-type output which is a differentiable function of its parameters. The operator product is then used instead of the minimum operator with a view to assessing the degree of activation of each rule, and the membership functions are constructed as a composition of differentiable functions.

Figure 9.1 reports the FNN in the case of two inputs, one output, and three membership functions in each premise. The structure of the rules achieved with this network is therefore:

$$R_i\text{: if } x_1 \text{ is } A_{j,1} \text{ and } x_2 \text{ is } A_{k,2} \text{ then } y \text{ is } y_i \tag{9.1}$$

where:

j,k = 1..3 indicates the membership function for each variable;
i = 1..9 indicates the i-th rule;
y_i is the i-th output, and $A_{j,h}$ is the j-th membership function associated with the hth output function (h=1,2).

The network output, effected with a combination of the outputs of each rule, is indicated with y^*.

As can be observed from Figure 9.1, the network topology is completely determined by the number of inputs and the number of membership functions for each input, and hence the *trial and error* phase is not needed for determining the number of neurons.

The number of membership functions must, however, be chosen for every variable. he network layers can be looked on as cascade stages, each one corresponding to a step of the fuzzy inference. In particular, the part of the antecedents is processed in the layers from A to C. The output of each neuron of layer D represents the degree of activation of each fuzzy rule. The remaining layers process the consequent part of the rules and calculate the crisp output value.

The relations describing the forward phase of the network are reported below:

$$O^A_{h,s} = wc_{h,s} + x_h \tag{9.2}$$

$$O^B_{h,s} = \frac{1}{1+\exp(-wg_{h,s} O^A_{h,s})} \tag{9.3}$$

$$O^C_{h,1} = O^B_{h,1} \tag{9.4}$$

$$O^C_{h,2} = O^B_{h,2} - O^B_{h,3} \tag{9.5}$$

$$O^C_{h,3} = O^B_{h,4} \tag{9.6}$$

$$O^D_i = \frac{O^C_{1,p} O^C_{2,q}}{\sum_{i=1}^{9} O^D_i} \tag{9.7}$$

$$O^E = y^* = \sum_{i=1}^{9} y_i O^D_i \tag{9.8}$$

where h = 1,2; s = 1....4; p = ...3; q = 1...3; and where O^L indicates the L-th output layer.

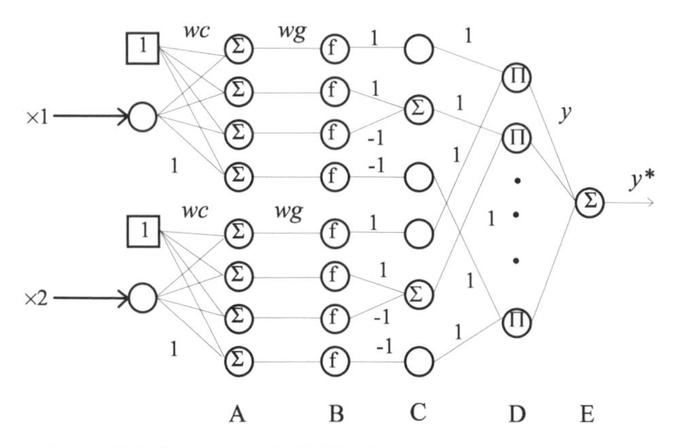

Figure 9.1. Structure of a FNN whose consequents are constant

As can be observed from the reported relation, the weights *wc* and *wg* determine the shape of membership function and the allocation in the universe of discourse, while the weight y_i represents the crisp output value of every rule. By appropriately choosing the weights *wc* and *wg*, the membership function can be initialized as reported in Figure 9.2, where the *medium* membership function is composed of the union of two sigmoidal functions, as can also be seen in equation 9.5.

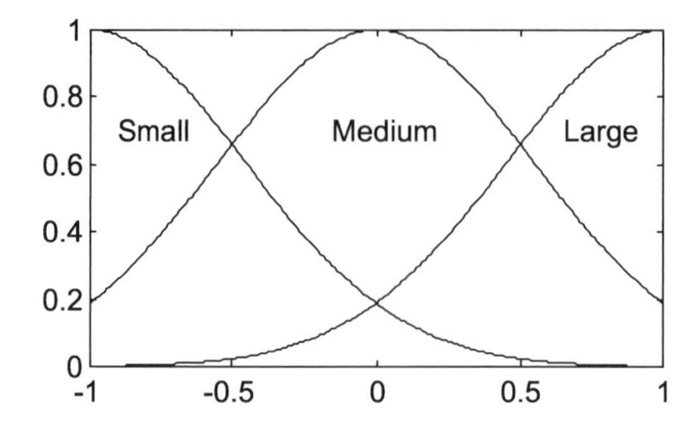

Figure 9.2. Initial configuration of the membership function

The considered structure can easily be extended to cover a generic number of inputs and membership functions. Even if more complex structures, which realize different structures for the consequents of each rule, may seem more suitable for facing some problems or for obtaining better performances, some considerations should be made. First of all, it can in fact be proved [3] that, under certain conditions, the simplified topology of Figure 9.1 produces fuzzy inference systems that can approximate with an *arbitrary degree of approximation* any non-linear function whatsoever defined on a compact set.

Despite that being true also for other network topologies, they often require a hard training phase, from the standpoint of computation, due to the complexity of the defuzzification process; generally, that does not correspond with improved performance for the network.

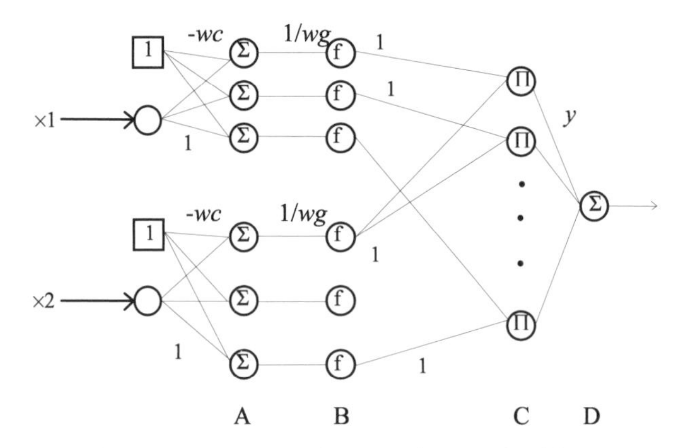

Figure 9.3. FNN structure with a Gaussian membership function

A further modification consists in using *Gaussian-type* membership functions instead of sigmoidal compositions. In that case, the FNN structure may be represented as in Figure 9.3.

The relations describing the forward propagation phase of the FNN in question are reported below:

$$O^A_{h,s} = x_h - wc_{h,s} \tag{9.9}$$

$$O^B_{h,s} = \exp\left(-\left(\frac{O^A_{h,s}}{wg_{h,s}}\right)^2\right) \tag{9.10}$$

$$O^C_i = \frac{O^B_{1,s} O^B_{2,p}}{\sum_{i=1}^{9} O^C_i} \tag{9.11}$$

$$O^D = y^* = \sum_{i=1}^{9} y_i O^C_i \tag{9.12}$$

where $h = 1,2$; $s = 1,2,3$; $i = 1...9$; $p = 1,2,3$. In this case, too, O^L indicates the output of the L-th layer.

9.3 Learning Algorithms for Neuro-fuzzy Networks

The learning algorithms for both FNNs proposed may be derived from a suitable modification of the classical back-propagation algorithm developed for the MLP [8]. The quadratic performance index that will be minimized is defined as:

$$E = \frac{1}{2}\sum_{i}^{N}(d_i - O_i)^2 \tag{9.13}$$

where d_i is the desired value of the i-th output O_i of the layer F.

The weights are modified so as to minimize the back-propagation error and the derivatives are calculated by applying the *chain rule*. The general formula for the weight adaptation algorithm may be written as:

$$w(t+1) = w(t) + \varepsilon \frac{\partial E}{\partial w} \tag{9.14}$$

where ε is the speed of learning.

9.4 Example: Interpolating a Non-linear Map with a Fuzzy Model

Let us consider the following non-linear function in the space x, y, z already seen in the example shown in Section 4.5.1:

$$z = xe^{(x-y)} + y\sin(2\pi x) \tag{9.15}$$

Let us suppose we wish to effect the interpolation operation described in Section 4.5.1 using a neuro-fuzzy-type of architecture [10]. The program used is called AFM version 2.0 [11] and also allows us to obtain FNNs with triangular-shaped membership functions.

The input variables (x, y) are both defined in the interval [-1, 1] and produce values of the variable z in the interval [-2, 8]. Let us also presume that the membership functions used in the interpolation procedure are triangular in shape. The generation of patterns is achieved in a completely analogous way to that seen in Section 4.5.1.

The validity of the model constructed is tested by carrying out a simulation on different patterns from those used during the learning phase. If the model is judged to be of low quality, the learning is repeated inserting more patterns or modifying the number of membership functions or the shape.

In the case in question, the model is structured in the following way:

Variable x:	*5 membership functions*
Variable y:	*2 membership functions*
Number of rules	*10.*

The membership functions are represented in Figures 9.4, 9.5, and 9.6.

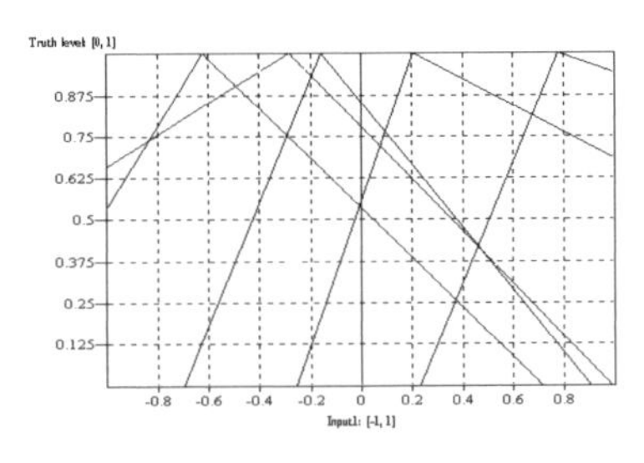

Figure 9.4. Variable x membership functions

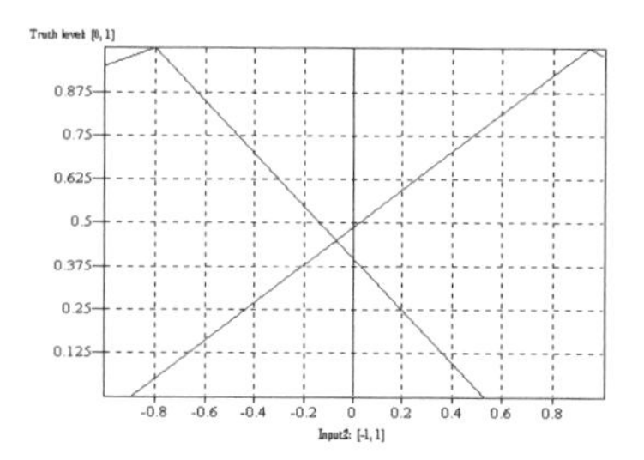

Figure 9.5. Variable *y* membership functions

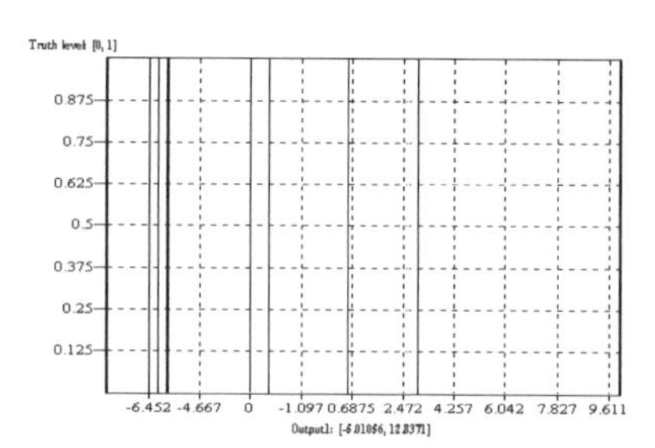

Figure. 9.6. Consequents of *z*

The rules of the model obtained are shown in Figure 9.7:

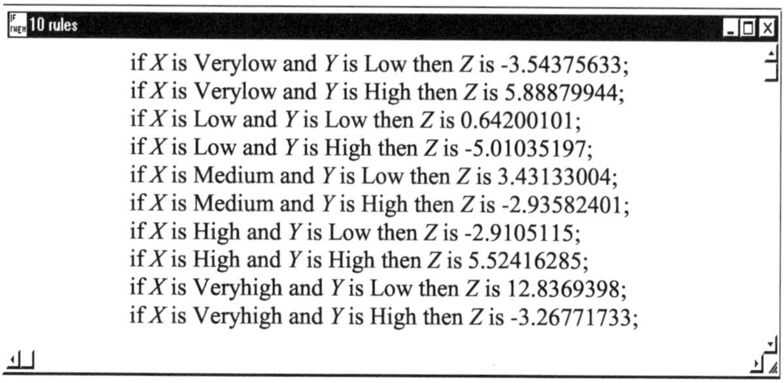

10 rules

if *X* is Verylow and *Y* is Low then *Z* is -3.54375633;
if *X* is Verylow and *Y* is High then *Z* is 5.88879944;
if *X* is Low and *Y* is Low then *Z* is 0.64200101;
if *X* is Low and *Y* is High then *Z* is -5.01035197;
if *X* is Medium and *Y* is Low then *Z* is 3.43133004;
if *X* is Medium and *Y* is High then *Z* is -2.93582401;
if *X* is High and *Y* is Low then *Z* is -2.9105115;
if *X* is High and *Y* is High then *Z* is 5.52416285;
if *X* is Veryhigh and *Y* is Low then *Z* is 12.8369398;
if *X* is Veryhigh and *Y* is High then *Z* is -3.26771733;

Figure 9.7. Fuzzy rules

Finally, Figures 9.8 and 9.9 show, respectively, the true values maps interpolated with the fuzzy model obtained. The values of the variables *x*, *y* are uniformly spaced over a grid in the interval [-1, 1] with a 0.1 step.

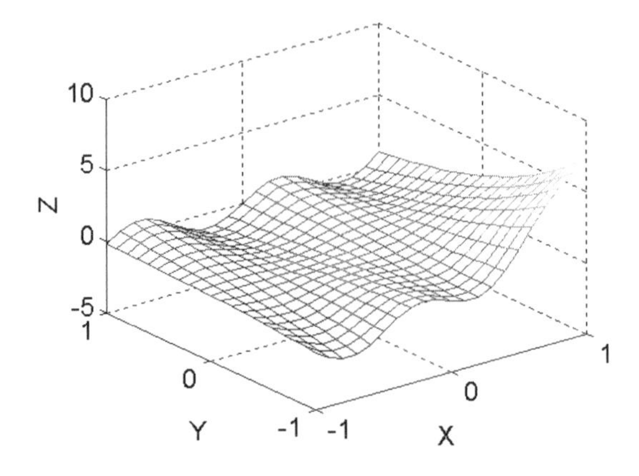

Figure 9.8. True values map

From Figure 9.10 it can be seen that the results are satisfactory. However, by comparing the results with those obtained from the interpolation shown in Section 4.5.1, it can be seen how the latter are better (by about one order of magnitude) in that the mean square error is:

MSE = 17.15*10-2

It can be noted that the mean error is close to zero and, to be precise:

ME = 11.48*10-3

which testifies to the fact that the interpolation is not polarized.

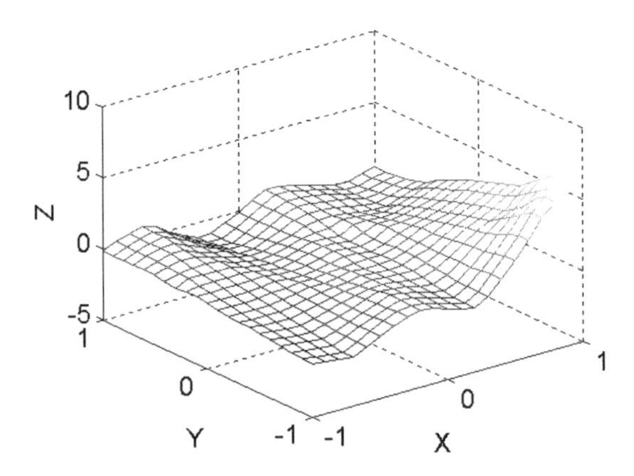

Figure 9.9. Interpolated values map

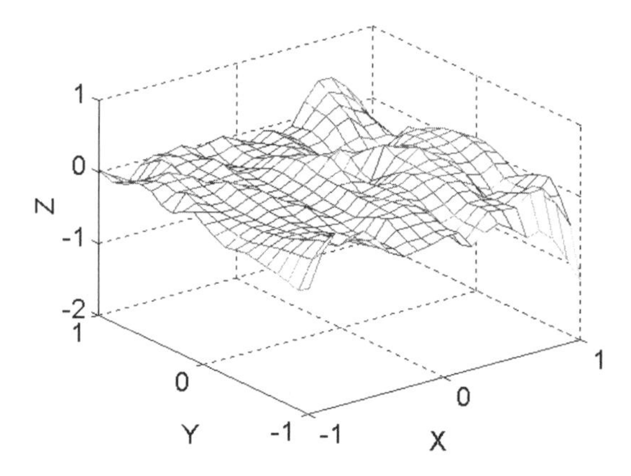

Figure 9.10. Difference map

9.5 Conclusions

The neuro-fuzzy structures described in this chapter, while being only part of those proposed in the literature, can be widely used. Quite apart from the structure used, it is useful to repeat once more the advantages afforded by these tools with respect to classical neural networks and the other optimization algorithms for fuzzy systems. In particular, it should be observed that, with respect to neural networks, neuro-fuzzy ones have two fundamental advantages: the possibility of using heuristic knowledge of the system to determine the number of fuzzy sets, and the possibility of choices based on heuristic knowledge of the system and the initial position of the fuzzy sets, thereby improving the convergence of the learning algorithm.

With respect to the other optimization methods used for fuzzy systems, neuro-fuzzy networks generally offer a lesser computational burden, although allowing us to modify with ease both the number of fuzzy sets and their position during the learning phase.

It should also be borne in mind that also for the neuro-fuzzy networks the learning can be effected by using evolutionist algorithms.

9.6 References

1. Shing J, Jang R. ANFIS: Adaptative Network Based Fuzzy Inference System. IEEE Trans. On System Man and Cybernetics. 1993; 23: 3
2. Baglio S, Fortuna L, Xibilia MG, Zuccarini P. Neuro-Fuzzy to Predict Urban Traffic. Proc. EUFIT94, Aachen, Germany, September 1994

3. Horikawa S, Furuhushi T, Uchikawa Y. On Fuzzy-Modelling Using Fuzzy Neural Networks with the Back-Propagation Algorithm. IEEE Transaction on Neural Networks. 1992; 3: 5
4. Gupta, M M, Qi J. On Fuzzy Neuron Models. Proc. Int. Joint Conf Neural Networks, Seattle, WA 1991; II: 431-436
5. Kosko B. Neural Networks and Fuzzy Systems: A Dinamical Systems Approach to Machine Intelligence. Prentice-Hall, Englewood Cliffs, NJ, 1992
6. Kosko B. Fuzzy Systems as Universal Approximators.Proc. IEEE Int. Conf. Fuzzy Syst., San Diego 1992; 1153-1162
7. Pedrycz W. Fuzzy Neural Networks and Neurocomputations. Fuzzy Sets Syst. 1993; 56: 1-28
8. Chin-Teng Ling, George Lee C S. Neural Fuzzy Systems. Prentice Hall PTR, Upper Saddle River, NJ 07458
9. Jang J.-S R., Sun C.-T, Mizutani E. Neuro-Fuzzy and Soft Computing. Matlab Curriculum Series
10. Nauck D, Klawonn F, Kruse R. Foundations of Neuro-Fuzzy Systems. John Wiley & Sons, Chichester, UK, 1997
11. AFM2 Ver. 2.0, S.G.S Thomson Microelectronics, 14 March 1997

10. Fuzzy Cellular Neural Networks

10.1 Introduction

In this, as in the previous chapter, we will deal with information processing systems that were originally inspired by the concepts underlying soft computing. The integration of fuzzy logic concepts in a widely spread architecture such as that of cellular neural networks in fact led to the birth of *fuzzy CNNs*.

From the moment of their introduction, cellular neural networks proved to be very powerful computation structures. At the same time, they should still be considered not easily and not immediately utilizable by the general kind of user. It has been proved [3] that they are as universal as the *Turing machine* whose programmability, given a precise task, depends on the choice of a *suitable template*. The lack of any one systematic method for determining the characteristic CNN parameters (templates) for a specific task remains one of the main factors limiting their widespread use. In Chapter 7, we presented an automatic extraction method for templates based on evolutionist optimization methods; instead, in the present chapter, we will describe a new computation structure which unites the characteristics of CNNs as widespread locally interconnected processing units together with the peculiarity of fuzzy logic to describe processes by sets of linguistic rules. In this new structure, called fuzzy cellular systems (FCS), the problem of template determination is shifted to the identification, often dictated by a knowledge of the specific phenomenon, of two sets of fuzzy rules, the first describing the *evolution* of an individual cell and the second effecting the *relationships* between the *locally interconnected cells*. We will show that also a FCS proves to be a *universal computation* structure whose peculiarity lies in its being able to solve reaction-diffusion (RD) types of differential equations. These equations govern all those complex phenomena, such as self-organization, formation of *Turing patterns*, and propagation of *auto-waves* in active media, which are displayed both at a microscopic and macroscopic level in numerous processes of physics, chemistry, biology, economics, sociology, medicine, and meteorology.

10.2 The Fuzzy Cellular System

Definition 10.1. *A fuzzy cellular system (FCS) is an array of M*N elementary fuzzy cells, all identical with each other and mutually interacting. a variable* $x_{ij}(k)$ *representing the state of the cell C(i,j) at a time k, is associated with each cell.*

Definition 10.2. *Each cell reacts with all the others in its neighborhood. Let* Nr_{ij} *be the r-the neighborhood of the generic cell C(i,j);* Nr_{ij} *is defined by the set:*

$$\{C(k,l) \mid \max(|k-i|,|l-j|) \leq r\}.$$

Definition 10.3. *The interactions of a generic cell C(i,j) with the cells in its neighborhood* Nr_{ij} *are determined by a set R1 of fuzzy rules.*

R1 calculates the state $x_{ij}(k)$ *of the cell C(i,j) taking into account the state* $x_{kl}(k)$ *of each cell of the neighborhood* Nr_{ij}.

Definition 10.4. *The dynamic state of an individual cell is determined by a set R2 of fuzzy rules. R2 calculates the state* $x_{ij}(k+1)$, *given* $x_{ij}(k)$ *and* $Nr_{ij}(k)$. *The dynamics of each cell is guaranteed by the set* $(R1 \cup R2)$, *whereas the overall behavior of the FCS is given by the mutual interactions between the M*N cells of the whole array, by means of the set R2.*

For example, if $X_{ij}(t)$ is the state of the generic cell C_{ij}, the dynamic evolution could be represented, for set $R1$, by rules of the type:

IF $X_{ij}(k)$ IS low AND $Nr_{ij}(k)$ IS high THEN $X_{ij}(k+1)$ IS small_positive,

and for set $R2$ of the type:

IF $X_{(i-1)j}(k)$ IS low AND $X_{i(j+1)}(k)$ IS medium AND ... THEN $Nr_{ij}(k)$ IS high

Given that at the edges of the cells there is not a complete neighbourhood, they are treated differently.

Three solutions are usually adopted.

a. *Zero flux conditions (Neumann)*: it is assumed that the missing cells always have a zero value state.
b. *Fixed potential conditions (Dirichlet)*: the cells at the borders always have a pre-established state.
c. *Toroidal conditions*: each of the border cells is connected with cells on the opposite border.

The fuzzy rules parameters must be determined according to the particular application. The basic diagram for a FCS composed of 3×3 cells arranged in a 2D matrix is shown in Figure 10.1.

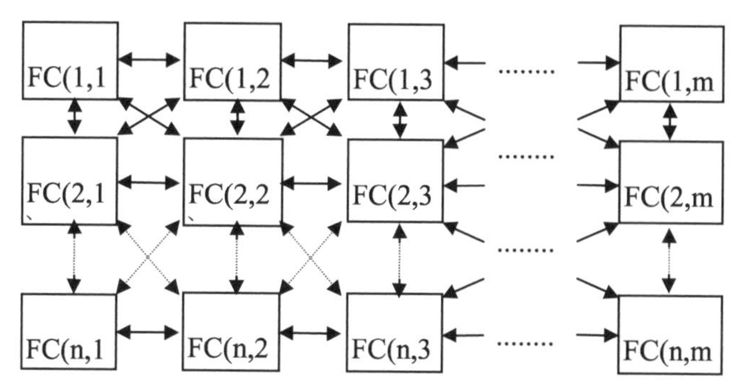

Figure 10.1. FCS composed of 3×3 cells arranged in a 2D array

10.3 Fuzzy Cellular Systems as Universal Computational Structures

Let us prove the calculation power of a fuzzy cellular system. The reasoning is based on the theorem of the *Turing universal machine* (TUM) which, in one of its many versions, can be synthesized thus: there exists no algorithm executed by a processor, which cannot be executed by a Turing machine.

In 1982, Berlekamp, Conway and Guy [4] demonstrated that any kind of architecture capable of implementing the game of life algorithm satisfies the Turing theorem, and therefore is equivalent to a TUM.

This finding was exploited by Chua, Roska and Venetianer [3] to prove that the architecture of a cellular neural network, the CNN universal machine (CNN-UM) is the equivalent of a Turing universal machine.

The *game of life* algorithm is a first order *cellular automata* which can be ideally played on an infinite two-dimensional matrix of cells. The initial condition consists in a random distribution of cells with only two possible states: *live c*ell - *dead cell*; the cells evolve in the subsequent instants of time (they are made discrete) according to the following rules:

1. if the generic cell $C(i,j)$ is alive at the instant k and its neighborhood with radius equal to 1 contains more than three live cells, then $C(i,j)$ will be dead at the instant k+1 (*dead because of overcrowding*);
2. if $C(i,j)$ is alive at the instant k and its neighborhood with radius equal to 1 contains less than two live cells, the $C(i,j)$ will be dead at the instant k+1 (*dead because of isolation*);
3. if $C(i,j)$ is dead at the instant k and its neighborhood with radius equal to 1 contains three live cells, the $C(i,j)$ will be alive at the instant k+1 (*birth*);
4. if $C(i,j)$ is alive at the instant k and its neighborhood with radius equal to 1 contains two or three live cells, then $C(i,j)$ will remain alive at the instant k (*survival*).

Starting from a random initial condition and after a sufficient number of moves, it can be immediately verified that the game of life evolves into a very rich and complex dynamic behavior: up to 15 different dynamic situations (*patterns*) can be observed.

Each *pattern* represents a particular configuration of cells: static cells, cells that oscillate with respect to a stable central position, or cells that move within the whole two-dimensional space until they interact with another pattern and destroy its dynamics (*glider gun pattern*). The equivalence with the Turing machine allows us, given the initial state, to deal with the predictability of the future state of the cell matrix in the same way as we solve the problem of stopping a Turing automaton program.

In order to extend these results to the fuzzy cellular system, we will give some preliminary definitions and assumptions so that the FCS structure may implement Rules 1 through 4 of the game of life.

Definition 10.5. *The basic unit of a fuzzy cellular system that implements a fuzzy game of life is a fuzzy cell made up of a continuous state variable $x_{ij}(t)$ which represents the individual state of health in the i-th position and j-th column of the matrix of cells. The state +1 corresponds to a live cell with excellent conditions of health, the state 0.5 corresponds to a live cell with normal conditions of health, while the state 0 corresponds to a dead cell; all the intermediate values between [0,1] are possible and represent different states of health.*

Assumption 10.1. A set of *R*1 rules. *Let us assume that the state of a generic cell is determined by means of an inference from the state of the cells from the neighborhood with radius equal to 1: in particular, a predominance of neighboring live cells with a good state of health will cause a worsening of the state of health of the cell in question owing to overcrowding. The same will happen if there is a predominance of live cells in a bad state of health; in this case, the worsening is due to their being exposed to contagion.*

When the neighboring cells, but not of a high number, have generally normal or good conditions of health, then a condition of improvement will arise in the state of health of the cell, which will increase the probability of its survival. These qualitative behaviors can be described by means of a set of fuzzy rules that make up the R1 set. For R1, the following relations hold:

$$Nr_y(t) = \sum_{(k,l)=i-1,j-1}^{i=1,j=1} x_{kl}(t)$$

$$\mu_k = (x_{ij}(t) \text{ is } A_k(x_{ij}(t))) \text{ and } (Nr_{ij} \text{ is } B_k(Nr_{ij}(t))) \tag{10.1}$$

$$R1\_out = \frac{\sum_{k=1}^{nrules} \mu_k \cdot FS_k}{\sum_{k=1}^{nrules} \mu_k}$$

where A_k and B_k are, respectively, the fuzzy sets of the state of the cell and the state of its neighborhood; k represents the degree of activation of the k-th fuzzy rule of the R1 set, and R1_out is a numerical value that can be interpreted as the "chance of survival" of the cell in question.

Assumption 10.2. A set of R2 rules. *With regard to the evolution of the state of each cell, the following considerations hold. Each cell will be considered dead if its chance of survival, i.e., its new state of health calculated by R1, is less than a fixed minimum threshold. On the contrary, when the new state of health calculated by R1 is greater than a fixed maximum threshold, there will be the birth of a new cell. In other cases, the state of the cell will be modified according to the following equation:*

$$x_{ij}(t+1) = x_{ij}(t) + \Delta x \cdot (R1\_out - 0.5) \tag{10.2}$$

Figures 10.2-3 describe the fuzzy game of life, while Figures 10.4-7 give some results.

The main difference from the traditional game of life is the introduction of a *non-binary state of health*, and consequently the introduction of the maximum and minimum thresholds and the parameter x. The choice of these parameters affects the dynamic behavior of FCSs but not the configuration of the final patterns nor the capacity to calculate.

Extending the results of Chua *et al.* [3] [4] and [5], it can be stated that FCS are *universal computation structures*, like the Turing automata, and therefore FCS structures can implement any algorithm whatsoever, quite independently of its complexity.

Finally, it should be borne in mind that to program a FCS means specifying the rules that govern the evolution and interaction of each elementary unit of calculation (the cell), and that such operations can often benefit from an expert human operator's direct knowledge of the process in question, thereby considerably simplifying the programming.

An example of a *hardware architecture* that implements a FCS is reported in [6].

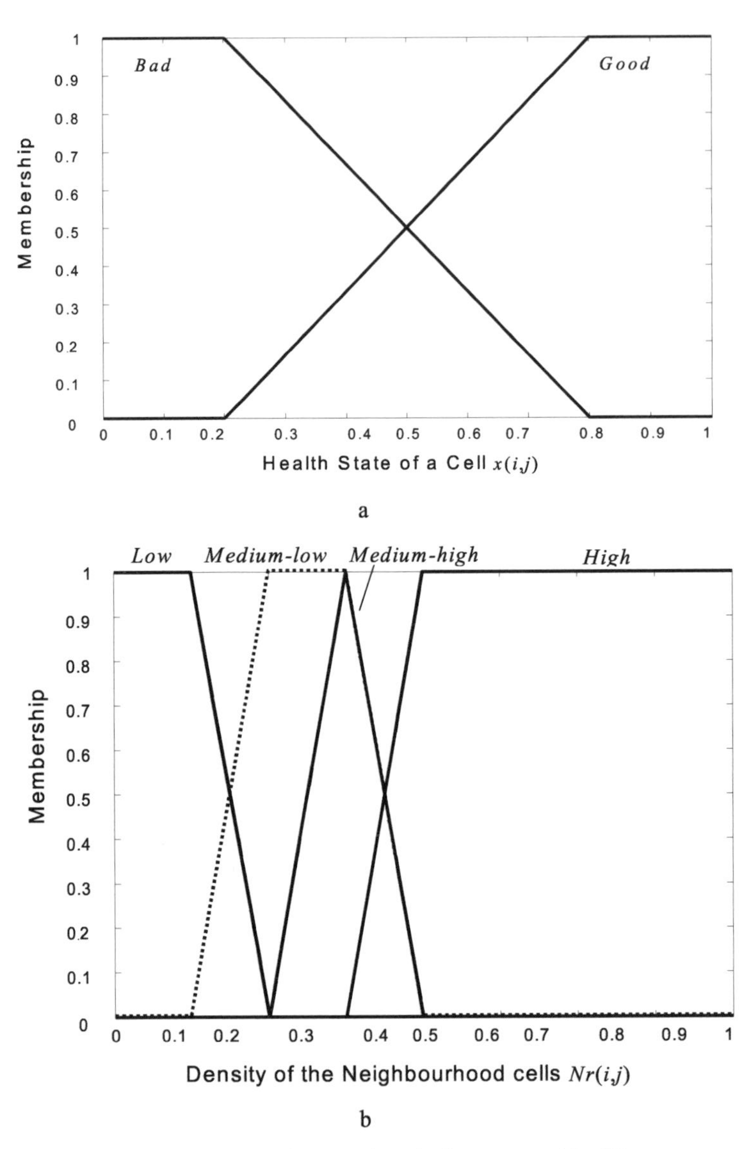

Figure 10.2. Membership functions in the IF part of the *R*1 set

***R*1 set**

Rules nr.1 (dead by overcrowding):
IF $X_{ij}(t)$ is Good AND $N_r(i,j)$ is High THEN FS_1 is 0.0

Rules nr.2 (dead by exposure):
IF $X_{ij}(t)$ is Good AND $N_r(i,j)$ is Low THEN FS_2 is 0.0

Rules nr.3 (birth):
IF $X_{ij}(t)$ is Bad AND $N_r(i,j)$ is Medium-High THEN FS_3 is 1.0

Rules nr.4 (survival):
IF $X_{ij}(t)$ is Whatever AND $N_r(i,j)$ is Medium-Low THEN FS_4 is 0.5

Figure 10.3. Rules in the THEN part of the R1set

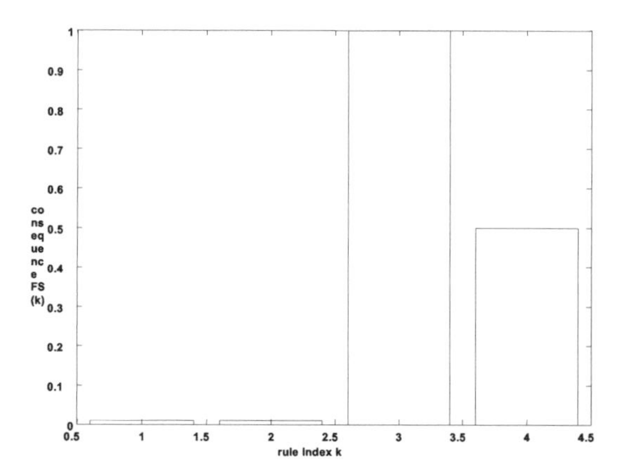

Figure 10.4. Parameters in the THEN part of the *R*1 set

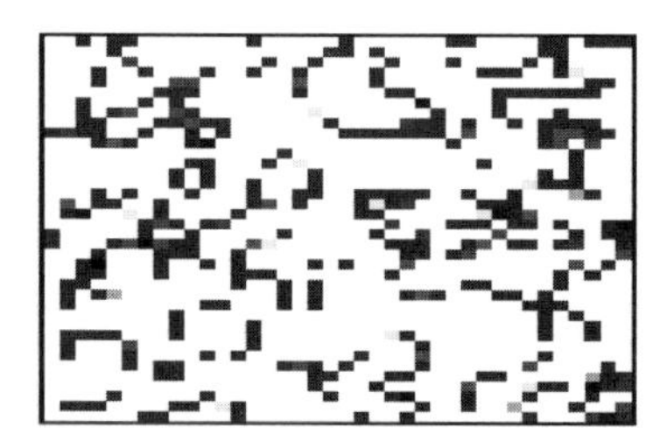

Figure 10.5. Results of the fuzzy game of life after three steps

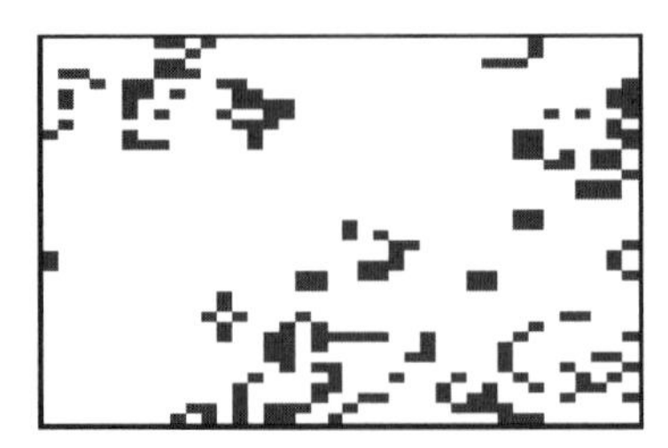

Figure 10.6. Results of the fuzzy game of life after 100 steps

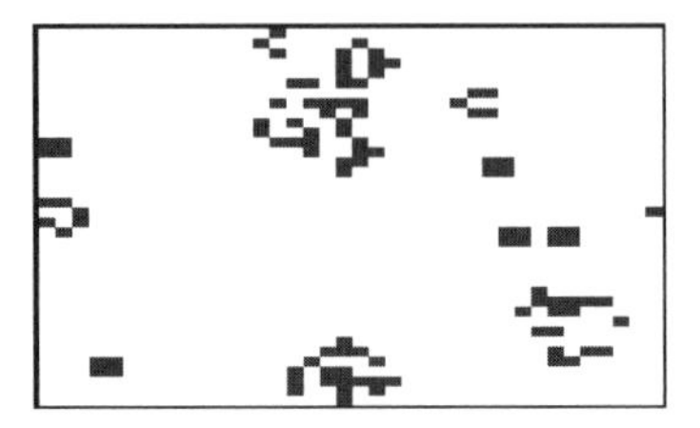

Figure 10.7. Results of the fuzzy game of life after 200 steps

10.4 FCS Applications in Image Processing

A *digital image* in black and white is represented by a matrix of real numbers representing the pixel gray levels making up the image [7]. Let us assume that these values of gray lie between -1 (white) and +1 (black). This matrix is used as an initial condition for the state variables of a FCS.
Let us also presume that the FCS never has an oscillating behavior and that it always reaches a steady state equilibrium. However, in other applications this condition will be removed. Therefore, starting from the initial pixel matrix, a FCS evolves by means of fuzzy inferences and reaches a condition of equilibrium. This condition represents the processed image. We will now give some examples of filters implemented by a FCS.

10.4.1 Noisy Corrupted Image Filtering

In this study, a *pixel* is considered a noise if its value of gray is *markedly different* from all the values of gray of the pixels belonging to its neighborhood. Let us consider two possible approaches. The first is based on the traditional *mean spatial filter method* [8].

Let $n_{ij}(\mathbf{x})$ be the neighboring function which measures the mean value of the state variables of the cells adjacent to the pixel C_{ij}, arranged above it, beneath it, to the left and to the right, as shown in Figure10.8; $n_{ij}(\mathbf{x})$ is given by the relation:

$$n_{ij}(\mathbf{x}) = \left(x_{i-1,j}(k) + x_{i+1,j}(k) + x_{i,j-1}(k) + x_{i,j+1}(k)\right) \cdot 0.25 \tag{10.3}$$

This function and the state $x_{ij}(k)$ of the cell C_{ij} are the two inputs of the set of rules determining the evolution of each cell C_{ij}. Let us consider two linguistic variables: *cell*, which represents $x_{ij}(k)$, and *n*, which represents $n_{ij}(\mathbf{x})$. The set of rules for the dynamic evolution of C_{ij} is given by:

IF *cell* IS *W* AND *n* IS *W* THEN *out* IS *N*

IF *cell* IS *B* AND *n* IS *B* THEN *out* IS *P* (10.4)

IF *cell* IS *W* AND *n* IS *B* THEN *out* IS *SP*

IF *cell* IS *B* AND *n* IS *W* THEN *out* IS *SN*

where *W* is the membership function for white, *B* is the membership function for black, *P* is the membership function for the positive values, *N* is the membership value for the negative values, *SP* is the membership function for the small-positive values, *SN* is the membership function for the small-negative values, and *out* represents the increase of $x_{ij}(k)$, to calculate $x_{ij}(k+1)$.

	$C_{I-1,j}$	
$C_{I,j-1}$	C_{ij}	$C_{I,j+1}$
	$C_{I+1,j}$	

Figure 10.8. Neighborhood of the pixel *Cij*

The membership functions are reported in Figure 10.9.

In agreement with the definition given for the scale of grays, the universes of discourse are between [-1,1].

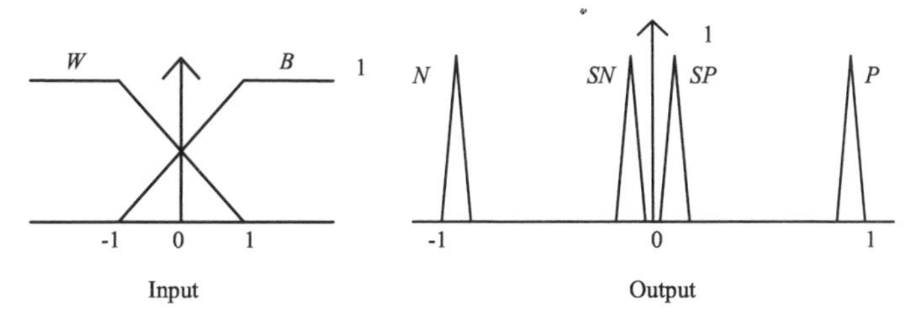

Figure 10.9. Membership Functions

For the centroid positions of the consequents membership functions, the following premises are assumed. The membership *P* has its centroid in +1, *N* has its centroid in -1, while for *NP* and *SP* the centroids are chosen in such a way as to guarantee that if the adjacent pixels have the same values of gray, then they will not have to be modified. For example, let us calculate the centroid *X* of *SP*, supposing that *cell* = *-1* (white pixel) and of the four neighboring pixels three are white and one is black; then *n* = (1+1+1-1)/4=0.5. Let us now calculate the activation values: *cell* is *W*, with an activation value of 1 (hence that of *B* will be 0) and *n* is *W* with an activation value of ¼ (and thus *B* will have a value of ¾). Under these conditions, from the set of rules (10.4), we obtain:

$$\min(1,1/4)*(-1)= - 1/4$$
$$\min(0,3/4)*(1)= 0$$
$$\min(1,3/4)*(X)= 3/4*X$$
$$\min(0,1/4)*(-X)= 0$$

and from this we obtain $out = (-1/4+0+3/4*X+0)$.

Given that we want this configuration of pixels to be maintained, the increase out must be negative, and thus $X < 1/3$. The value of *SP* is therefore determined by following this *trial and error* procedure, while for symmetry the centroid of *NP* will be fixed at $-X$.

Let us consider the image of Figure 10.10, made up of 38×29 pixels. This image is disturbed by adding impulsive and Gaussian noise which will modify 20% of the pixels value. The *impulsive noise* is formed by spurious black and white pixels, whereas the Gaussian noise has spurious pixels of different values of gray [7]. Figures 10.11 and 10.12 report, respectively, the disturbed images and the filtered image. In the case of Gaussian noise, the image processed by FCS is identical with the original one of Figure 10.12b, whereas in the case of impulsive noise the slightly less good results are shown in Figure 10.12a.

We can say that the method of filtering the spatial mean works better with Gaussian-type noise.

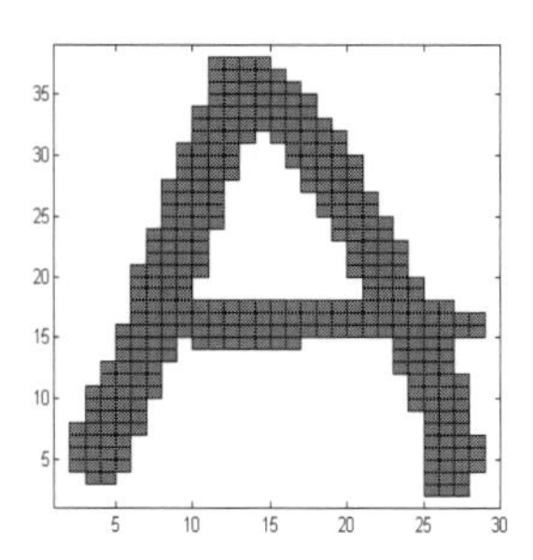

Figure 10.10. Original image

To improve the *filtering capacity* with respect to *impulsive noise*, we can introduce a different approach based on the concept of median functions for calculating $n_{ij}(\mathbf{x})$. The median *m* of a set of values is that value for which one half of the values of the set is smaller than *m* and the other half is larger than *m*. For example, 1 is the median value of the set (10, 0.5, 1, 1.5, 0.5). In this case, the set of pixels in the neighborhood of a cell includes the cell itself, and hence it is formed by the five cells of Figure 10.8.

The rules are the same as in the previous case, but the centroid positions of *SP* and *SN* can no longer be calculated as before since that would lead to the banal position of $X<1$. The centroid position choice is made on a *time-by-time* basis, calibrating the rules according to the type of filter desired. For impulsive noise, the membership functions are reported in Figure 10.13, while Figure 10.14 shows the result with the filter based on the median-based approach. Even if the filtered image is not exactly equal to the original, the results with the median-based approach are undoubtedly better than those based on the spatial mean method.

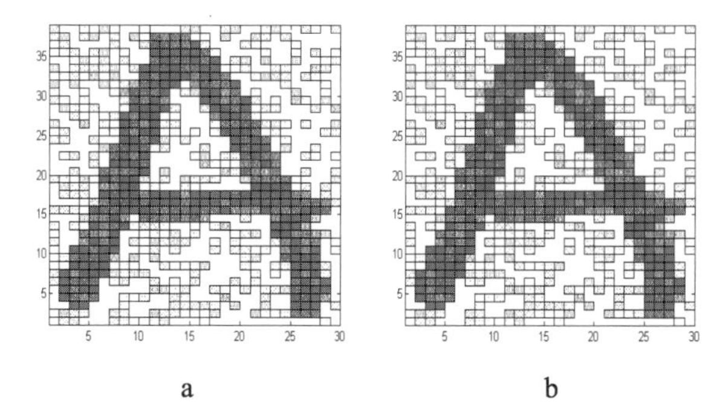

Figure 10.11. a. image disturbed by impulsive noise; b. image disturbed by Gaussian noise

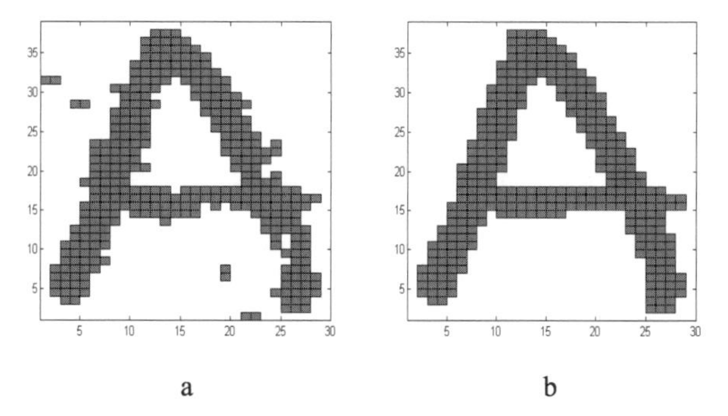

Figure 10.12. a. image disturbed by impulsive noise filtered using the spatial mean method; b. image disturbed by Gaussian noise filtered using the spatial mean method

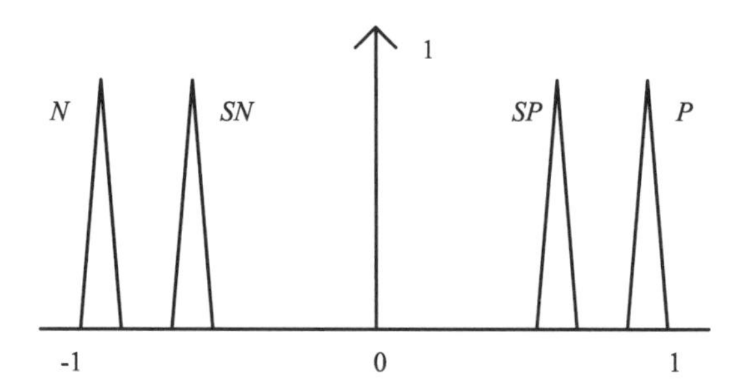

Figure 10.13. Membership functions for the consequents

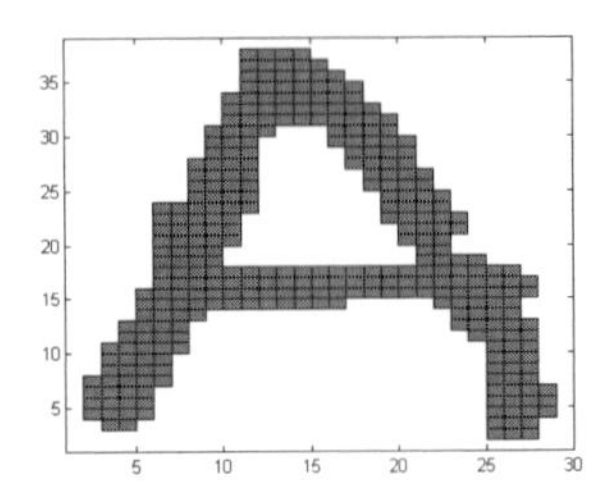

Figure 10.14. Image disturbed by impulsive noise filtered using the median method

10.5 Detecting Correlations between Characters

Detecting *correlations* between the pixels in an image (*connected components detection* (CCD)) is an extremely useful procedure in many algorithms for identifying characters [7]. In particular, a *horizontal CCD* consists in representing the continuous sequences of pixels having the same chromatic characteristics, belonging to a horizontal string. For example, for a row with three sequences of black pixels, the horizontal CCD operation is reported in Figure 10.15, where the sequence count is considered as starting from the right.

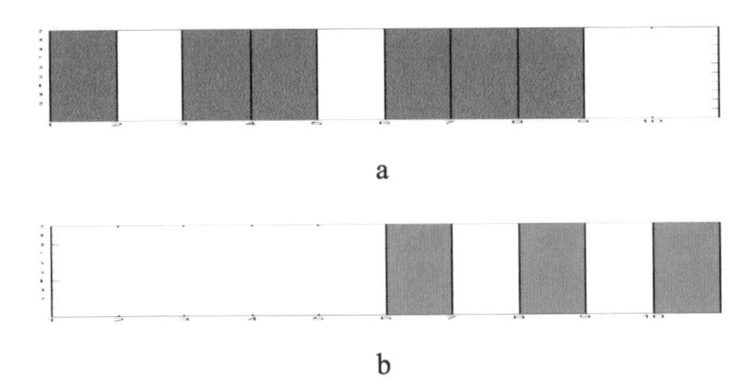

Figure 10.15. Horizontal CCD: a. original condition; b. processed image

Figure 10.16. Output membership functions

We will now give a possible implementation of the horizontal CCD algorithm by means of a FCS.

In this case, there are four input variables: *io* is the value of state $x_{ij}(k)$ of the cell in question C_{ij}, *p* is the state of the cell to the left of C_{ij}, s is the state of the cell to the right of C_{ij}, and *ss* is the value of the state $x_{i,j+2}(k)$.

The output membership functions are reported in Figure 10.16, where *B* represents the white chromatic values, *N* the black ones, and *V* represents the neutral values; in addition, *V* also accounts for all those cases where no cells exist adjacent to the cell in question (*border cells*).

The linguistic description of the CCD algorithm satisfies the following fuzzy rules:

IF *p* IS *B* AND *io* IS *B* AND *s* IS *B* THEN *out* IS *B*
IF *p* IS *B* AND *io* IS *B* AND *s* IS *N* THEN *out* IS *B*
IF *p* IS *B* AND *io* IS *N* AND *s* IS *B* AND *ss* IS *B* THEN *out* IS *B*
IF *p* IS *B* AND *io* IS *N* AND *s* IS *B* AND *ss* IS *N* THEN *out* IS *N*
IF *p* IS *B* AND *io* IS *N* AND *s* IS *N* THEN *out* IS *B*
IF *p* IS *N* AND *io* IS *B* AND *s* IS *B* THEN *out* IS *N*
IF *p* IS *N* AND *io* IS *B* AND *s* IS *N* AND *ss* IS *B* THEN *out* IS *B*
IF *p* IS *N* AND *io* IS *B* AND *s* IS *N* AND *ss* IS *N* THEN *out* IS *N*
IF *p* IS *N* AND *io* IS *N* AND *s* IS *B* THEN *out* IS *N*
IF *p* IS *N* AND *io* IS *N* AND *s* IS *N* THEN *out* IS *N*
IF *io* IS *N* AND *s* IS *V* THEN *out* IS *N*
IF *p* IS *N* AND *io* IS *B* AND *s* IS *V* THEN *out* IS *N*
IF *p* IS *B* AND *io* IS *B* AND *s* IS *V* THEN *out* IS *B*
IF *p* IS *B* AND *io* IS *N* AND *s* IS *B* AND *ss* IS *V* THEN *out* IS *B*
IF *p* IS *N* AND *io* IS *B* AND *s* IS *N* AND *ss* IS *V* THEN *out* IS *B*

The results of implementing the horizontal CCD algorithm with a FCS are reported in Figure 10.17.

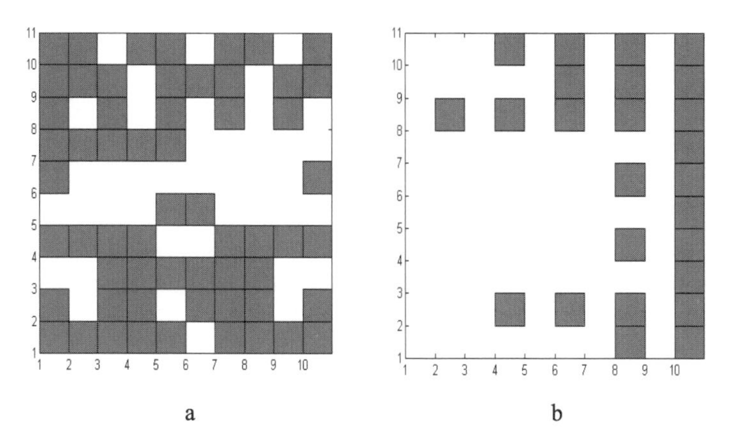

Figure 10.17. FCS for a horizontal CCD: a. initial condition; b. processed image

10.6 Reaction-diffusion Equations Solved by FCS

Turing's studies in the 1950s on *symmetry breaking* and *homogeneity* in the field of *morphogenesis* [8] were followed in the subsequent years by numerous other important studies aiming at analyzing complex phenomena which arise in biology, chemistry, ecology, and electronics [9], [10]. All these processes, due to interactions between simple units, similar one to another, are described by a set of differential equations with partial derivatives, known in the literature as *reaction & diffusion* (RD) equations:

$$\frac{\partial \mathbf{u}}{\partial t} = \mathbf{f}(\mathbf{u}) + \mathbf{D}\nabla^2 \mathbf{u} \tag{10.5}$$

where $\mathbf{u} \in \Re^n$ and $f(\bullet) \in \Re^n$ are, respectively, the state of an individual cell and a non-linear function; $\mathbf{D}=\{D_i\}$ are the diffusion coefficients, and $\nabla^2(\mathbf{u})$ is the Laplace operator on $\mathbf{u}$. For example, the 2-dimensional Laplace operator on $\mathbf{u}(x,y)=[u_1(x,y),\ldots, u_n(x,y)]$ is:

$$\nabla^2 u_i \triangleq \frac{\partial^2 u_i}{\partial x^2} + \frac{\partial^2 u_i}{\partial y^2} \qquad i = 1,2,\ldots,n \tag{10.6}$$

As discussed in Section 10.2, the dynamic behaviour of locally interconnected systems, as well as the formation of patterns and propagation of waves in active media, depends on suitable conditions at the boundary, which for the border cell $C(i,j)$, $i,j = 1, \ldots, N$ we can formalize thus:

a. *Dirichlet Conditions*
   $C(i,0)=E_1$, $C(i,N+1)=E_2$, $C(0,j)=E_3$, $C(N+1,j)=E_4$; with E_1, E_2, E_3, and E_4 being constant values;

b. *Neumann Conditions*
   $C(i,0) = C(i,1)$, $C(i,N+1) = C(i,N)$,
   $C(0,j) = C(1,j)$, $C(N+1,j) = C(N,j)$;

c. *Toroidal Conditions*
   $C(i,0)=C(i,N)$, $C(i,N+1)=C(i,1)$,
   $C(0,j)=C(N,j)$, $C(N+1,j)=C(1,j)$.

As was reported by Chua [10], some well-known RD equations, both 1D and 2D, can be implemented by means of CNN architecture. Examples are the Fisher equation in genetics, the Brusselator equation in thermodynamics, and the analyses of some processes in ecology and biology. The CNN approach is based on a circuital adaptation which allows us to solve numerically and quickly all those processes described in terms of RD equations. However, it is not at all simple to determine the RD equation for a process whose dynamic behavior derives from the nature of the local interconnections of the individual elements. Thus, it appeared quite natural that we should exploit the FCS computational potential in order to emulate fuzzy reaction and diffusion phenomena, such as pattern generation, spiral wave formation, auto-wave propagation in active media, and self-organization phenomena.

10.7 Pattern Generation

To employ the same terminology as Turing introduced for biological processes, the complex dynamics due to reaction and diffusion phenomena are generated by the continual interaction between a catalyzing factor (*catalytic morphogene*) and an inhibitor (*inhibitory morphogene*).

Locally, the catalyzer reacts with itself, increasing the degree of excitation, but also activating the inhibitor. At the same time, a long-range inhibition mechanism is set up, produced by a differentiated diffusion: the catalytic morphogene has a smaller diffusion power with respect to the inhibitory one ($D_{INHIBITORY} \approx 50{*}D_{CATALYTIC}$). When the reaction phenomenon is compensated by such a diffusion phenomenon, starting from random initial conditions and suitable boundary conditions, the process, which is made up of very many interacting agents, produces regular dynamic behaviors called patterns. Let us now consider the following hybrid (analytic + fuzzy) non-linear system:

$$\frac{\partial U_{i,j}}{\partial t} = \gamma \cdot \left(F1\left(U_{i,j}\right) \cdot U_{i,j} + f_v \cdot V_{i,j}\right) + D_u \cdot F2\left(U_{i,j}, N_r\left(U_{i,j}\right)\right)$$

$$\frac{\partial V_{i,j}}{\partial t} = \gamma \cdot \left(g_u \cdot U_{i,j} + g_v \cdot V_{i,j}\right) + D_v \cdot F2\left(V_{i,j}, N_r\left(V_{i,j}\right)\right) \tag{10.7}$$

$$Nr\left(C_{i,j}\right) = \frac{1}{4}\left(C_{i-1,j} + C_{i+1,j} + C_{i,j-1} + C_{i,j+1}\right)$$

where $F1(.)$ and $F2(.)$ are the fuzzy sets of the FCS whose membership functions are presented in Figure 10.18a and b and are defined by the set of rules:

Definition of $F1(x)$:

IF (x IS Low) THEN $F1$ is Negative
IF (x IS Medium) THEN $F1$ is Zero
IF (x IS High) THEN $F1$ is Positive

Definition of $F2(x.Nr)$ (Fuzzy Laplacian):

IF (x IS Low) AND (Nr is Low) THEN $F2$ IS Zero
IF (x IS Low) AND (Nr is Medium) THEN $F2$ IS Medium-Pos
IF (x IS Low) AND (Nr is High) THEN $F2$ IS Positive
IF (x IS Medium) AND (Nr is Low) THEN $F2$ IS Medium-Neg

IF (x IS Medium) AND (Nr is Medium) THEN $F2$ IS Zero
IF (x IS Medium) AND (Nr is High) THEN $F2$ IS Medium-Pos
IF (x IS High) AND (Nr is Low) THEN $F2$ IS Negative
IF (x IS High) AND (Nr is Medium) THEN $F2$ IS Medium-Neg
IF (x IS High) AND (Nr is High) THEN $F2$ IS Zero

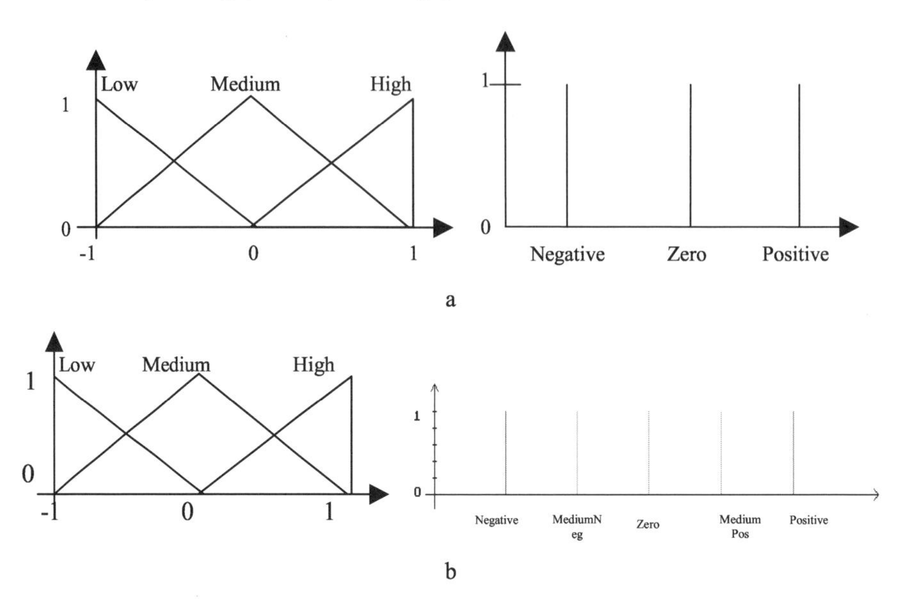

Figure 10.18. a. output membership and crisp functions used for pattern formation to define F1(x); b. Output membership and crisp functions used for pattern formation to define F2(x,Nr)

The remaining parameters have the following constant values:

$f_v=-1;\ g_u=0.1;\ g_v=-0.2;\ \gamma=10;\ D_u=0.2;\ D_v=40$

Figure 10.19 shows the pattern formation process starting from random initial conditions with toroidal boundary conditions.

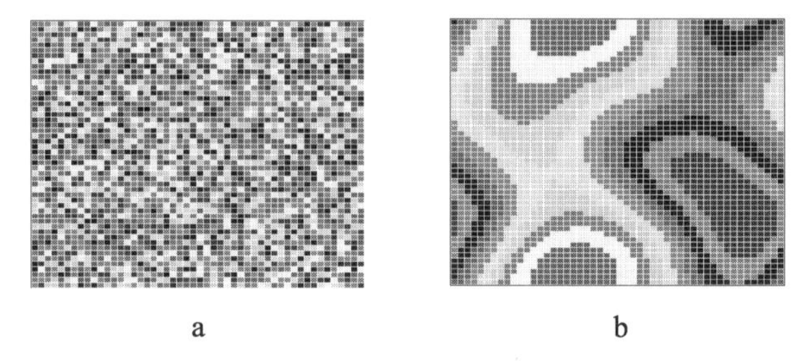

Figure 10.19. Pattern formation in fuzzy cellular systems: a. initial conditions; b. steady state (after 2100 steps)

10.8 Formation and Propagation of Spiral Waves

Spiral waves are complex phenomena, generated by non-linear cellular systems under particular initial conditions. Spiral waves are considered to be an excellent model for describing the phases of arrhythmia (infarct) in the heart: when subjected to particular solicitation, the heart beats irregularly, generating spiral waves that lead to defective pumping of the blood.

Let us consider the following hybrid set:

$$\begin{aligned}
\frac{\partial U_{i,j}}{\partial t} &= -U_{i,j} + (1+\gamma)\cdot F1(U_{i,j}) - F1(V_{i,j}) + D_u \cdot F2(F1(U_{i,j}), N_r(F1(U_{i,j}))) + i_1 \\
\frac{\partial V_{i,j}}{\partial t} &= -V_{i,j} + (1+\gamma)\cdot F1(V_{i,j}) + F1(U_{i,j}) + D_v \cdot F2(F1(V_{i,j}), N_r(F1(V_{i,j}))) + i_2 \\
Nr(C_{i,j}) &= \frac{1}{4}\left(C_{i-1,j} + C_{i+1,j} + C_{i,j-1} + C_{i,j+1}\right)
\end{aligned} \tag{10.8}$$

where $F1(.)$ and $F2(.)$ are the same fuzzy sets as those reported in Figure 10.18, while the constant parameters have the following values: γ=0.7, D_u=D_v=0.1, i_2=-i_1=0.3.

This kind of FCS evolves by generating a single spiral wave in Figure 10.20, while Figure 10.21 shows the collision of two waves; finally, Figure 10.22 reports the generation of a vortex due to the collision of four spiral waves.

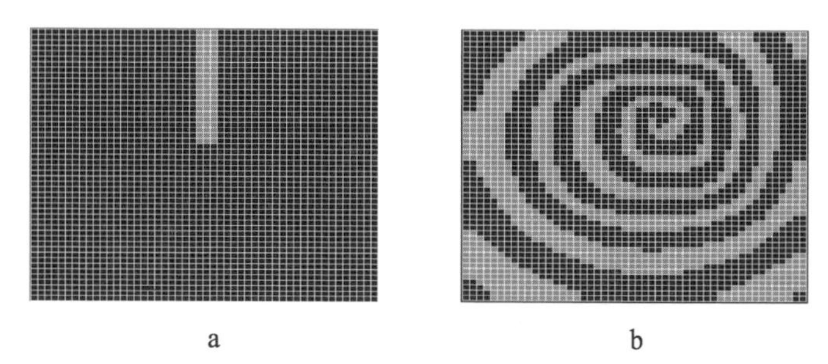

a b

Figure 10.20. Spiral wave generation in a FCS: a. initial condition; b. steady state (after 500 steps)

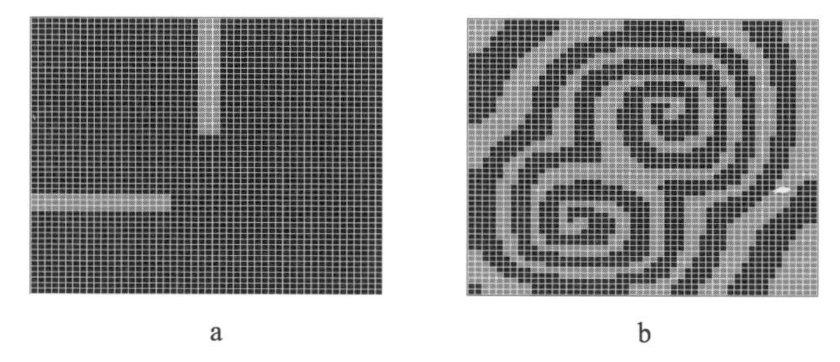

a b

Figure 10.21. Collision of two spiral waves in a FCS: a. initial condition; b. steady state (after 500 steps)

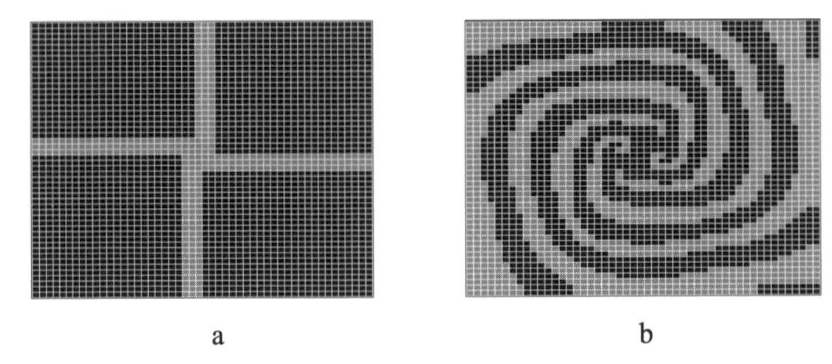

Figure 10.22. Generation of a vortex: a. initial condition; b. steady state (after 500 steps)

From the applications reported the versatility of fuzzy CNNs can be readily understood, especially for describing complex phenomena. Further stimuli and greater detail can be found in [6] and [10-13].

10.9 References

1. Chua LO, Yang L. Cellular Neural Networks: Theory. In Trans. on Circuits and Systems. 1988; 35: 1257-72
2. Chua LO., Yang L., Cellular Neural Networks: Applications.In Trans. on Circuits and Systems. 1988; 35: 1273-90
3. Chua LO., Roska T, Venetianer PL. The CNN is Universal as the Turing Machine In IEEE Trans. on Circuits and Systems I. 1993; 40: 4
4. Berlekamp E, Conway JH, Guy RK. Winning ways for your mathematical plays. In NY Academic. 1982; 2:25: 817-850
5. Crounse KR, Chua LO. The CNN Universal Machine is as Universal as the Turing Machine.In IEEE Trans. on Circuits and Systems I. 1996; 43: 4
6. Caponetto R, Lavorgna M, Occhipinti L. Fuzzy Cellular Systems: Characteristics and Architecture. In Fuzzy Hardware, editor. Kluwer Ac. Ed. 1997; Chapter 14
7. Gonzalez RC, Woods RE. Digital Image Processing. Addison Wesley. 1992
8. Turing AM. The Chemical Basis of Morphogenesis. In Phil. Trans. Roy. Soc. Lond. 1952; 237 (B.641): 37-72
9. Prigogine I, Lefever R. Symmetry Breaking Instabilities in Dissipative Systems, II. In J. Chem. Phys. 1968; 48: 4: 1695-1700
10. Chua LO, Hasler M, Moschytz GS, Neirynck J. Autonomous Cellular Neural Networks: A Unified Paradigm for Pattern Formation and Active Wave Propagation. In IEEE Trans. on CAS-I. 1995; 42: (10): 559-77
11. Baglio S, Fortuna L, Manganaro G. Fuzzy Cellular Systems for a New Paradigm of Computation. In Engineering Applications of Artificial Intelligence Journal, Pergamon Press 1997; 10: (1): 47-52
12. Caponetto R, Fortuna L, Lavorgna M, Occhipinti L. Fuzzy Cellular System for Image Processing. In WILF '97; Bari , 1997
13. Caponetto R, Occhipinti L, Lavorgna M, Fortuna L, Di Bernardo G. Cellular Fuzzy Processor: A New Architecture to Explore Complexity in Locally Interconnected Systems. In IEEE Int. Conf. On Electronics Circuits and Sistems. Lisbona, September, 1998

11. Fuzzy Systems Optimization by Means of Genetic Algorithms

11.1 Introduction

In the previous chapters, addressed to the theoretical introduction and applications of fuzzy control, we always referred to the ability of an expert, capable of transforming into linguistic form the actions to be applied to the system.

However, an expert is not always at hand to describe the functioning of the system, and indeed in the majority of cases the set of rules is tentatively determined and, generally speaking, the system must be optimized.

The present chapter aims to introduce the use of genetic algorithms for automatic design and optimization of a set of fuzzy rules.

Several approaches have been proposed in literature and the research being carried out makes the subject one of the most vital among soft computing methodologies [4-7].

In this chapter two examples will be reported; the first one regarding the fuzzy control of temperature in a greenhouse, and the second one concerning with the optimal design of fuzzy filters for processing signals.

11.2 Controller Representation and Objective Function

The parameters to be optimized in a fuzzy controller are related to the membership functions of the fuzzy input and output sets, both regarding shape and their position in the universe of discourse, and to the number of rules employed, which need not necessarily be equal to the combination of all the input variable memberships.

What acquires great importance is the choice of an objective function that will guarantee good performance for the system, which is formed by the process and the closed-loop fuzzy controller. Determining the fitness function is subjective and clearly connected to the application in question.

In many cases, the choice is made to minimize any error between the system's output and the desired one, but it is all the same possible to consider performance indices which allow for other factors, such as the energy spent for the control, economic aspects, and so on.

The first problem to be faced, however, is the choice and codification of the membership functions.

Below, we will report two types of membership function, Gaussian and triangular, where different approaches are required for their codification.

If the membership function is Gaussian, as in Figure 11.1, we need to determine the two values of a and b which fix the mean value and the variance; if instead it is triangular, as in Figure 11.2, the position of the three apices must be fixed. In this case, the number of parameters is three and it drops to two for symmetrical memberships.

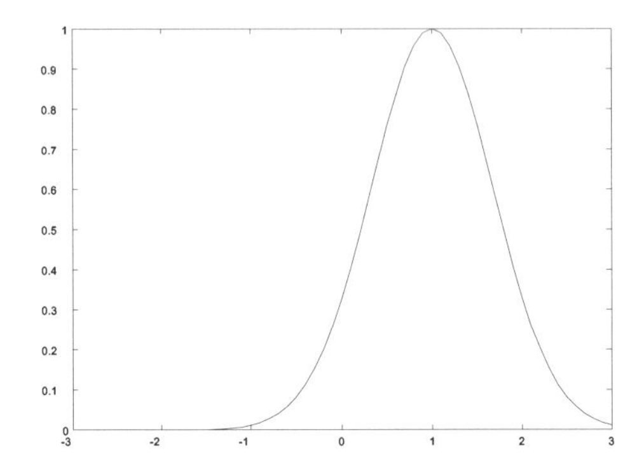

Figure 11.1. Gaussian membership and the parameters (a and b) to be optimized

The codification of a phenotype for Gaussian membership functions can be carried out using a string of bits, and thus an element of the population assumes the following form:

(10110011001001.....10101010111010)

| a_1 | b_1 | | a_n | b_n |

or else, when the codification is by means of real numbers, the form will be: (a_1b_1...........a_nb_n)

The generic fuzzy rule R_i of the system in question assumes the form:

R_i: If x_1 *is* A_i and x_2 *is* B_i and ... and x_n *is* T_i then out_i *is* Z_i

In this case, we must optimize the values characterizing the A_i, B_i, ...,Z_i membership functions of the antecedents and consequents. If the consequents are crisp values and not fuzzy sets, as in the following rule:

R_i: If x_1 *is* A_i and x_2 *is* B_i and ... and x_n *is* T_i then out_i *is* y_i

then we will have to optimize the membership functions A_i,.B_i, ...,T_i of the antecedents and the numerical y_i values of the outputs. In the light of that, a set of fuzzy rules, the so-called fuzzy system, can be defined by means of a multiparameter concatenation of genotypes representing all the memberships defined for the input and output variables.

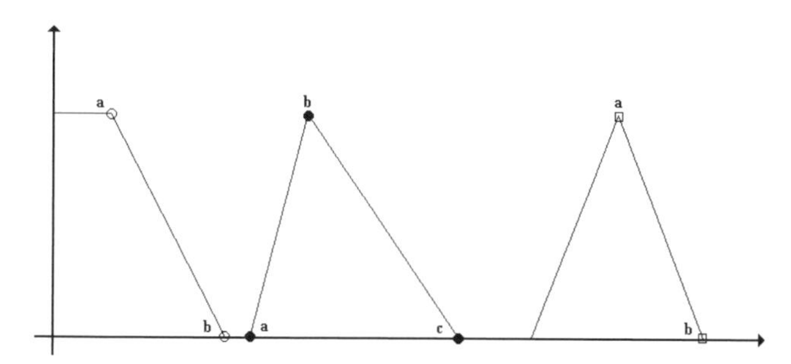

Figure 11.2. Triangular membership function and parameters (a, b, and c) to be optimised

One possible representation is therefore the one below.
Let us give the following notation:

- n: number of input variables;
- in_k membership, k=1, 2, ..., n: number of fuzzy sets defined for the k-th input variable;
- **o**: number of output variables;
- out_h membership, h=1, 2, ..., **o**: number of fuzzy sets defined for the h-th output variable.

The individual element of the population now assumes the following form:

$M\_in_{1,1}$	...	$m\_in_{1,in1}$	...	$m\_in_{n,1}$	...	$m\_in_{n,inn}$
$M\_out_{1,1}$	...	$m\_out_{1,out1}$	...	$m\_out_{\mathbf{o},1}$	...	$m\_out_{\mathbf{o},outo}$

where every string represents, respectively:

n	...	n-th input
First output	...	o-th output

If the number of inputs is high, and the memberships defined on the individual variables are equally numerous, the number of rules may increase considerably. Again using genetic algorithms, also the number of rules may be optimized, thereby making the fuzzy system less heavy both from the standpoint of computation (fewer inferences) and of hardware (less memory used). In order to obtain that, one possible approach is, by means of the objective function, to assign a weight to every rule, Thus, by following the characteristic philosophy of genetic algorithms, only those strings with the number of optimal rules will be selected and combined with each other to generate an optimal fuzzy system.

11.3 Fuzzy Control of Greenhouse Temperature

An example regarding the approach proposed is given below. What we want to determine is an optimal fuzzy controller that will regulate the temperature inside a greenhouse. The number of inputs and outputs of the controller has been fixed by the expert. In particular, the inputs are two, the temperature and its variation, while the output is a quantity connected to the opening of the window of the greenhouse.

The SIMULINK® diagram used during the simulation is reported in Figure 11.3. The model of the greenhouse was constructed following what was reported in [1-2] and using the information supplied by the expert.

The fuzzy controller was implemented using the MATLAB® Fuzzy Toolbox, and the parameters were optimized by a genetic algorithm also written in MATLAB®. The parameters characterizing the controller were the following:

- input variables: error and error variation;
- 5 fuzzy sets on the universe of discourse of the error;
- 3 fuzzy sets on the universe of discourse of the error variation;
- 7 rules determined on the basis of the expert's knowledge.

The total number of variables regarding the individual element of the population is 23 and is made up of:

- 5 m.f. first input characterized by 2 parameters: 10 variables
- 3 m.f. second input characterized by 2 parameters: 6 variables
- 7 crisp output values: 7 variables

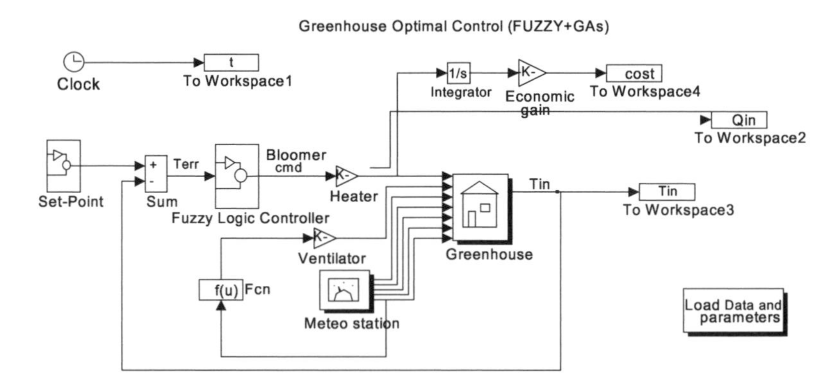

Figure 11.3. SIMULINK® diagram of the fuzzy control system for greenhouses

The rules used are reported in the following table:

Table 11.1. Rules for the greenhouse fuzzy controller

R_1	IF Error IS Pos THEN Blooming_control IS Mbf7
R_2	IF Error IS Zero AND Change_in_error IS Pos THEN Blooming_control IS Mbf2
R_3	IF Error IS Zero AND Change_in_error IS Zero THEN Blooming_control IS Mbf3
R_4	IF Error IS Zero AND Change_in_error IS Neg THEN Blooming_control IS Mbf4
R_5	IF Error IS Neg THEN Blooming_control IS Mbf0
R_6	IF Error IS PosLow THEN Blooming_control IS Mbf6
R_7	IF Error IS NegLow THEN Blooming_control IS Mbf1

The genetic algorithm parameters are the following:

8 bits for every variable
80 element population.
30 generations
Pcross=0.7
Pmut=0.003

The objective function for the *k*-th element of the population has the following form:

$$E(k)=sqrt(error^t*error)/nsamples$$

where error indicates the difference between the simulated output and the one imposed, and where *nsamples* is the number of samples (a number kept constant during all the optimization).

The results obtained are reported in Figures 11.4 and 11.5.

These results are interesting when compared with those obtained by means of classical control methods such as bang-bang control, which is usually used in these cases.

With regard to the two control methods, Figure 11.6 reports the following dimensions: reference temperature and controlled temperature, external temperature disturbance, amount of heat supplied and relative cost.

For bang-bang control, Figure 11.6a, one can note an internal temperature oscillation which is controlled with respect to the reference one. This error is practically annulled in the case of a fuzzy controller, Figure 11.6b. Another advantage of the fuzzy controller, compared with the traditional ones, is that of markedly reducing the amount of heat supplied, and thus the running costs.

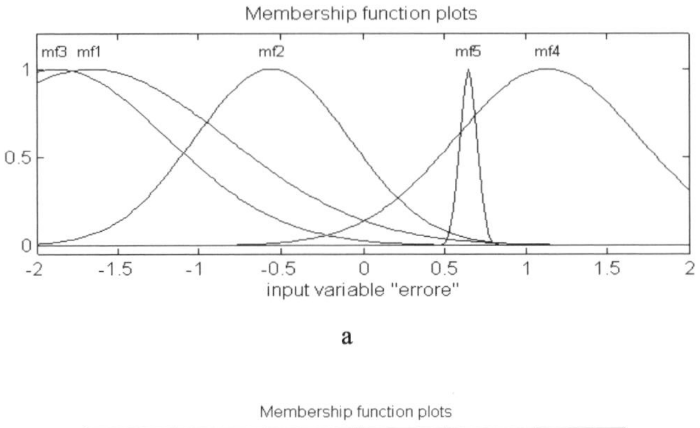

a

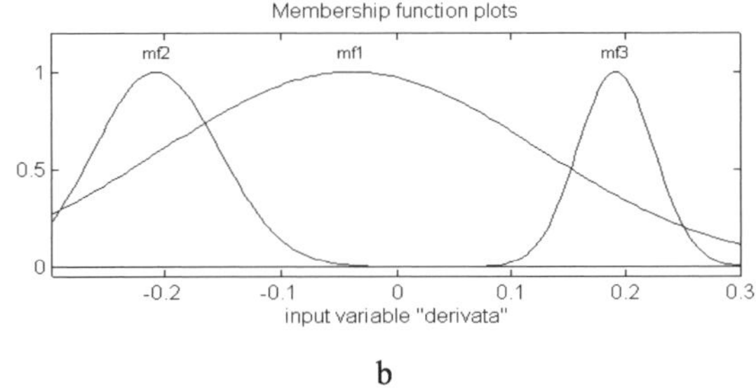

b

Figure 11.4. Optimized membership functions: a. antecedents; b. derivative

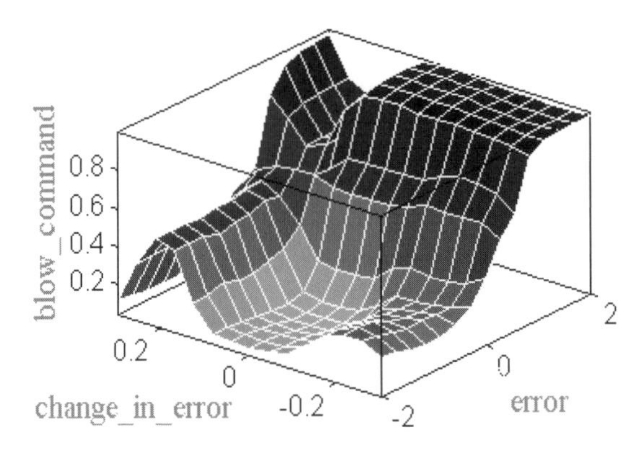

Figure 11.5. Control map determined by genetic algorithms

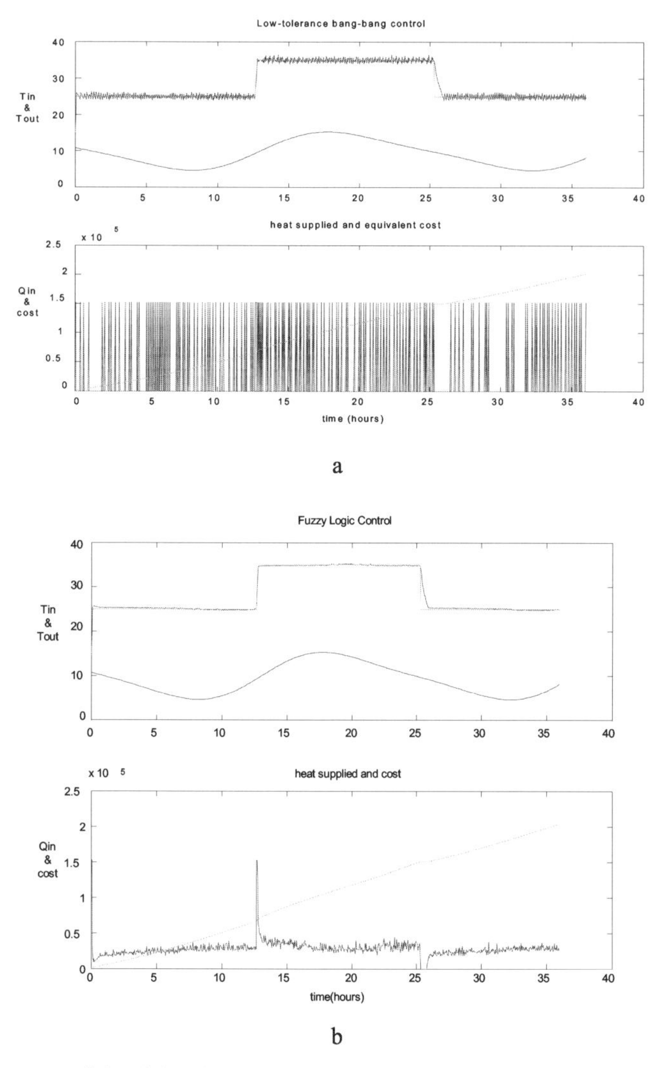

Figure 11.6. a. traditional threshold control ± 1 °C threshold; b. fuzzy control optimized by GAs

11.4 Optimal Realization of Fuzzy Filters

In its most general form, a filter is represented by the following finite difference equation:

$$y(k) = a_1 y(k-1) + a_2 y(k-2) + ... + a_n y(k-n) +$$

$$b_1 u(k) + b_2 u(k-2) + \dots + b_m u(k-m)$$

where a_i and b_j are the coefficients of the filter, and $y(k\text{-}i)$ and $u(k\text{-}m)$ are, respectively, the output and input samples at definite instants of time.

Therefore, assimilating a digital filter to a fuzzy system, the latter can be defined by a set of rules, each one having the following structure:

$$\text{if } u(k) \text{ is } A_0 \text{ and } \dots u(k-n) \text{ is } A_n \text{ and}$$

$$y(k-1) \text{ is } B_1 \text{ and } y(k-n) \text{ is } B_n \text{ then } y_i(k) = y_0$$

where $u(k\text{-}j)$, $y(k\text{-}h)$ with j=0,....,n; h=1,....,n are the input and output filters of the fuzzy filter, k is the sampling time, and A_j e B_h are fuzzy sets.

For the sake of simplicity, in this model at the output $y_i(k)$ of every rule we have assigned a crisp value rather than a fuzzy value. To calculate the output $y(k)$, given the m rules representing the filter, the centroid method was used, which calculates $y(k)$ as:

$$y(k) = \frac{\sum_{i=1}^{m} \mu(R_i(k)) * y_o^i}{\sum_{i=1}^{m} \mu(R_i(k))}$$

where $\mu(R_i(k))$ represents the degree of activation of the i-th rule at the instant k calculated by:

$$\mu(R_i(k)) = A_0^i u(k)) * \dots * A_n^i u(k-n)) * B_0^i y(k)) * \dots * B_n^i y(k-n))$$

with $A_j^i u(k)$ and $B_h^i u(k)$ representing the membership values of the variables $u(k\text{-}j)$ e $y(k\text{-}h)$ with respect to the fuzzy sets A_j^i and B_j^i, while m is the number of all the possible combinations of the fuzzy sets A_j^i and B_j^i. The filter defined in this way is an IIR filter. If it is intended to represent a FIR filter, it is sufficient to not consider the output samples $y(k\text{-}h)$, and so the generic rule assumes the following form:

$$\text{if } u(k) \text{ is } A_o^i \text{ and } u(k) \text{ is } A_n^i \text{ then } y_i(k) = y_o^i$$

Fuzzy filter design [3] is motivated by advantages arising from limited calculation times and the possibility of a hardware implementation with readily available fuzzy processors.

The methodologies reported in the literature for identifying fuzzy filters are based on heuristic techniques for neural modeling or for classical optimization methods.

In particular, in [3] the Sugeno method was employed. If this technique is adopted, to impose the specifications in the frequency domain, a set of input-output data must be used which is obtained from an equivalent filter. That involves two disadvantages: the need to create a fictitious set of input-output data and the presence of noise, which will inevitably be present in these data.

Here, we report a new approach to designing fuzzy filters which allows us to directly impose the specifications in the frequency domain, although the filter representation is maintained in the time domain. In this case, too, genetic algorithms have been used for the optimal choice of the parameters characterizing the filter.

11.4.1 Fuzzy Filters and Genetic Algorithms

As said above, the design of a fuzzy filter consists in determining the parameters which characterize the various rules. In particular, we must establish the form of the fuzzy sets and the values to assign to the outputs y_o^i, once the specifications of the filter have been fixed in the frequency domain.

So as not to make the algorithm excessively heavy, the number of fuzzy sets and their (Gaussian) form have been fixed a priori. Thus, the output values of the m rules, which will be identified later by the vector y_o^i, still remain to be determined. During the optimization procedure, the first step consists in fixing the mask desired for the filter response in the frequency domain. The cut-off frequency and passband , *i.e.*, the characteristic dimensions of the filter response, can therefore be fixed.

The parameters to be optimized are codified once again exploiting the multiparameter concatenation described in the chapter dealing with GAs.

The physical qualities of every element of the population must allow for the distance between the desired response and that obtained during the optimization process. Therefore, it is convenient to consider the square error of the differences between the two responses so as to minimize the distance between the two curves as the frequency varies.

Let us suppose we wish to design a low-pass filter, one possible choice for the fitness function might be the following:

$$F = \sum_{i=1}^{F_c} abs(\log(G(i))) + \sum_{i=F_c+1}^{F_s/2} abs(\log(abs(1 - G(i))))$$

where *Fc* indicates the cut-off frequency, *G(i)* the *i*-th sample of the filter frequency response module, and Fs the sampling frequency.

The fitness function as thus defined allows us to weight the error in a differentiated way for the higher and lower frequencies with respect to the cut-off one. During the optimization procedure, the fuzzy filter frequency response was determined using the cross-relation spectrum between the input noise and the corresponding output noise.

Given the symmetry of the filter response with respect to the output signal, the y_o^i vector parameters are antisymmetrical. This feature allows us to halve the number of parameters to be optimized.

Also in the case of fuzzy filters, if we intend to design high-order ones, we have to solve a multi-variable optimization problem. In fact, as the order increases, we must consider more I/O samples and consequently both the length of the rules and their number will rise. To deal with this problem, we can decompose the original filter into several lesser-order ones, in particular, when considering FIR filters, the decomposition will be of the serial kind, whereas if the filters are IIR ones, it will be of parallel type. We will report below an example to clarify this decomposition.

The following results were obtained exclusively considering FIR filters and the parameters characteristic of the GA are those reported in the following table:

Table 11.2. GA parameters used for designing a fuzzy FIR filter

8 bits for every variable
60 element population.
50 generations
Pcross=0.7
Pmut=0.0053

The frequency trend of the two second-order filters obtained using, respectively, the fuzzy approach and the traditional one are reported in Figure 11.7. For both filters, the cut-off frequency was fixed at 30 Hz, while the number of fuzzy filter rules was set equal to eight.

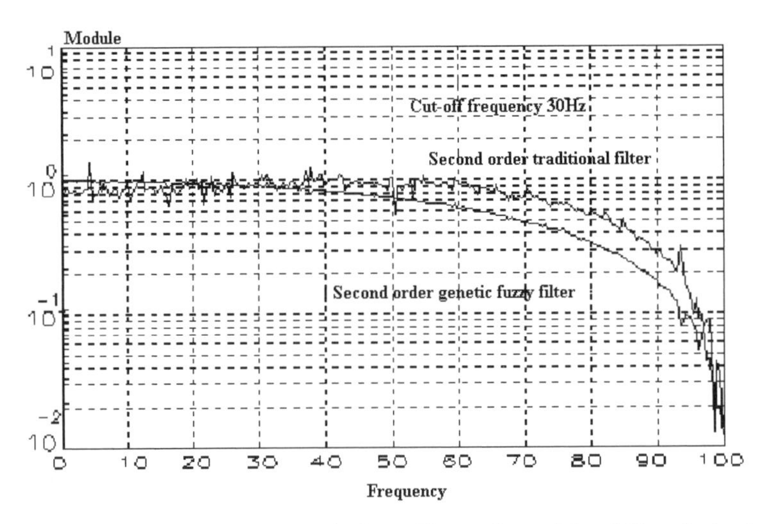

Figure 11.7. Module diagram of a genetic fuzzy filter and a traditional filter, both of second order (cut-off frequency: 30 Hz - second-order traditional filter - second-order genetic fuzzy filter - module - frequency)

It is evident that the fuzzy filter is more selective than the traditional one of the same order. Figure 11.8 reports an analogous comparison for two third-order filters. In this case, too, the fuzzy filter cuts off better at frequencies above 30 Hz.

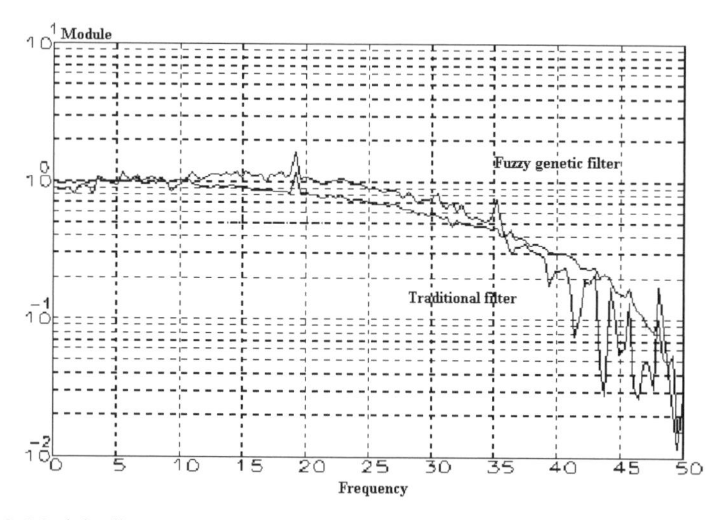

Figure 11.8. Module diagram of a genetic fuzzy filter and a traditional filter, both of third order

If we intend to realize a high-order filter, it is convenient, as already mentioned, to use the approach based on decomposition of the high-order filter into several filters of reduced order. The example in Figure 11.9 reports the comparison between two twelfth-order filters: the fuzzy filter obtained as a cascade of four third-order filters is the directly designed classic twelve-order filter.

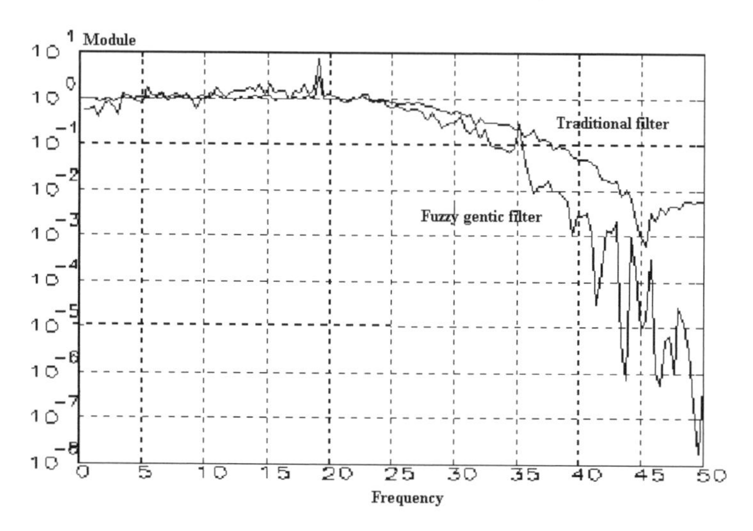

Figure 11.9. Module diagram of a genetic fuzzy filter and a traditional filter, both of twelfth order

Although, at varying frequency, the fuzzy filter module presents a less regular trend compared with the traditional filter, on the whole its response continues to display greater selectivity in the same order of conditions.

The approach named genetic-fuzzy for designing digital filters is a new conception and few references to it can be found in the literature. The encouraging results obtained and the ever greater interest shown in using fuzzy logic for digital signal processing make the proposed algorithm an interesting point of departure for further studies in depth.

11.5 References

1. Beccali, Giaccone A, Panno G. Modello di Calcolo per l'Analisi del Comportamento Termico delle Serre - Prima Parte. Energie Alternative HTE. July-August 1992; 4: 18: 283-289
2. Beccali, Giaccone A, Panno G. Modello di Calcolo per l'Analisi del Comportamento Termico delle Serre - Seconda Parte. Energie Alternative HTE. Sept. - Oct. 1992; 4: 19: 405-415
3. Baglio S, Fortuna L, Lo Presti M, Vinci C. Fuzzy IIR and FIR Filter', Proceeding of the 36th Midwest Conference on Circuit and Systems, Detroit, USA , August 1993
4. Caponetto R, Fortuna L, Nunnari G, Occhipinti L. Soft Computing Techniques on Automatic Synthesis of Greenhouse Climate Controllers. GALESIA '97
5. Caponetto R, Fortuna L, Muscato G, Nunnari G. Search of Optimal Realization Matrix for Filter Implementation by Using Genetic Algorithms. Proc of IEEE ISCAS Chicago, USA. May 3-6 1993
6. Caponetto R, Fortuna L, Vinci C. Design of Fuzzy Filters by Genetic Algorithms. IEEE ISCAS. London, May 30 - June2 1994
7. Caponetto R, Lavorgna M, Lo Presti M, Rizzotto GG. How GAs and Neuro-Fuzzy System can be Used for Automatic Controllers Design. IFSA 95. VI International Fuzzy System Association World Congress. San Paulo, Brazil, July 22-28 1995

12. Neuro-fuzzy Strategies for Monitoring Urban Traffic Noise

12.1 Introduction

This chapter deals with a case study regarding the problem of determining a model for *acoustic pollution* level assessment in urban areas. The model was identified by soft computing techniques based on the use of some parameters considered most relevant.

In particular, a classical empirical model, a fuzzy model, a neural model, and a neuro-fuzzy one were compared using the same set of experimental data during the model determination phase.

Acoustic pollution in urban areas degrades the quality of life and has thus aroused increasing attention on the part of government bodies and the scientific community [1-2].

The main source of noise in urban areas is the circulation of motor vehicles; the noise produced depends on certain physical parameters, such as, for example, the structure of the street and the height of the buildings flanking it.

The models available in the literature [3-6] generally hypothesize a *linear relation* between the inputs and outputs. These models arose from empirical and experimental considerations, and the results obtained are not satisfactory.

In this chapter, we will propose new models for predicting acoustic noise, based on soft computing techniques and in particular on MLP-type neural networks, with a set of fuzzy rules determined by the optimization method proposed by Sugeno and with a neuro-fuzzy network having constant consequents. The models were obtained using a set of measurements recorded in various urban roads of a medium-sized city.

12.2 The Set of Measurements

The external acoustic observations were carried out at sixteen stations sited in residential, commercial, and industrial areas.

The roads chosen can be grouped as follows:

- city-center streets (6 measurement sites);
- suburban roads (4 measurement sites).

For each measurement, a sound-level meter capable of recording 100 samples a second and a video recorder were used. From the data collected by the sound-level meter, the equivalent sound pressure level was obtained, defined as:

$$L_{eq} = 10\log\frac{1}{T}\sum 10^{\frac{L_i}{10}} \tag{12.1}$$

where:

T=3600 s observation time;
Li sound level (dB) measured in 1s.

The other parameters held necessary for determining the model were measured off-line using the video camera recordings, and are the following:

- the number of vehicles in circulation (1/h);
- the nature of the vehicles;
- the geometric section of the roads (width of road and height of buildings), (m);
- the type of roadbed.

In addition, the equivalent number of vehicles was defined:

$$n_{eq} = n_{cars} + c_1 * n_{mc} + c_2 * n_{hv} \tag{12.2}$$

where:

n_{cars} is the number of cars (1/h);
n_{mc} is the number of motorcycles (1/h);
n_{hv} is the number of heavy vehicles (1/h).

The values c_1=3 and c_2=6 are two coefficients recommended in the literature [12] which allow the number of motorcycles and heavy vehicles to be transformed into an equivalent number of cars, so as to produce the same L_{AeqT}. value. The model therefore assumes the following form:

$$L_{AeqT} = f(n_{eq}, h, w) \tag{12.3}$$

Table 12.1 reports the ranges of the parameters measured.

Table 12.1. Ranges of measured parameters

	L_{AeqT}	n_{eq}	h	w
Min	68.3	924	12	10
Max	81	10300	33.5	52
Average	76.06	3846	17.54	25.52

From Table 12.1, it can be observed that the city is very polluted. The L_{AeqT}. value never drops below 65 dB, a value considered by Italian law as the maximum limit for residential areas, and even reaches 81 dB, which is above the acceptable level. Figure 12.1 reports some values of the measured data.

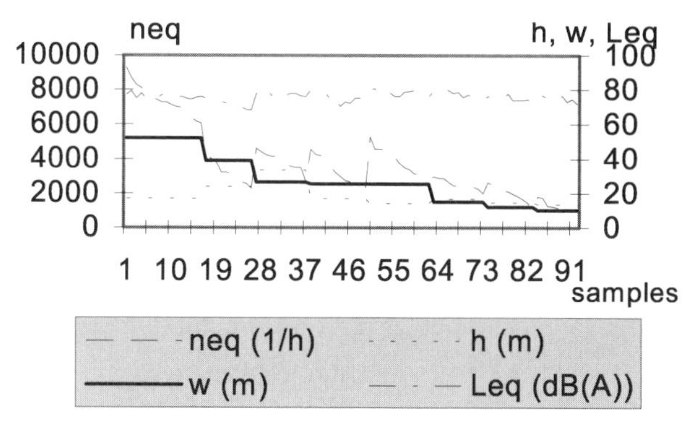

Figure 12.1. Sub-set of measured data

12.3 Classical Models

The models reported in the literature [3], [5] are linear models which assume the following form:

$$L_{AeqT} = a_0 + a_1 w + a_2 h + a_3 n_{eq} \ldots \tag{12.4}$$

Modelization efforts have been addressed to determining the $\boldsymbol{a}_i$ parameters by experimental considerations and optimization methods. The following model (suggested in [3]) is a linear function of the logarithm of some of the parameters considered and will be indicated below as the Josse model:

$$L_{AeqT} = 38.8 = 15\log(n_{eq}) - 10\log(w) \tag{12.5}$$

To obtain results in conformity with the experimental data, non-linear models should be adopted. The main problem is thus the choice of the particular non-linearity to take into consideration, *i.e.*, the form of the function $f(\bullet)$.

12.4 The Fuzzy Model

The form of fuzzy inference used for constructing the acoustic pollution model is the one proposed by Sugeno and described in Chapter 2 [7],[13], [16]:

$$\text{if } x_1 \text{ is } A_1 \text{ and } x_2 \text{ is } A_2 \text{ and} \ldots \text{ and } x_n \text{ is } A_n \quad \text{so } y=g(x_1,x_2,\ldots,x_n) \qquad (12.6)$$

In the case of the acoustic pollution model proposed here, the variables are the following:

- x_1: equivalent number of vehicles (1/h);
- x_2: average height of buildings (m);
- x_3: width of roads (m);
- y: equivalent level of sound pressure (L_{AeqT})(dB)

The steps followed for determining the *fuzzy model*, *i.e.*, for determining the fuzzy sets of the antecedents, the number of rules, and parameters of the consequents, were already described in Chapter 2. In particular, with regard to the choice of fuzzy sets, we started from the *M200* configuration in which 2 fuzzy sets are considered for the first variable. With the previously described procedure, the research tree shown in Figure 12.2 was obtained.

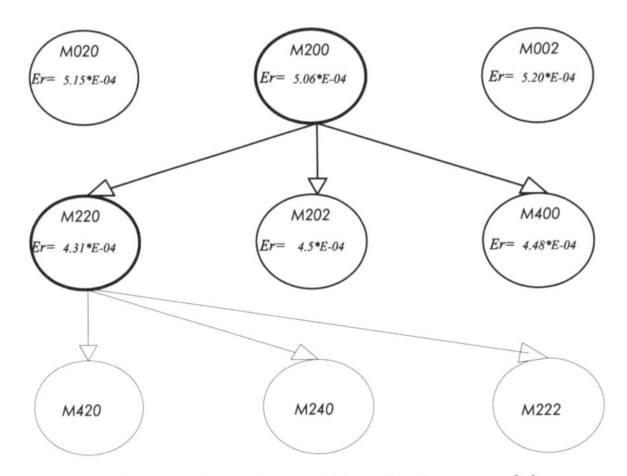

Figure 12.2. Research tree for fuzzy models

In Figure 12.2, every circle represents a model. The circles at the same level refer to models of the same complexity: that is to say the first level refers to models that contain two fuzzy sets in one variable, and so on. For every level, the circle with greatest thickness indicates those models which have the least error value of the models of the same complexity. These models were chosen for the next step.

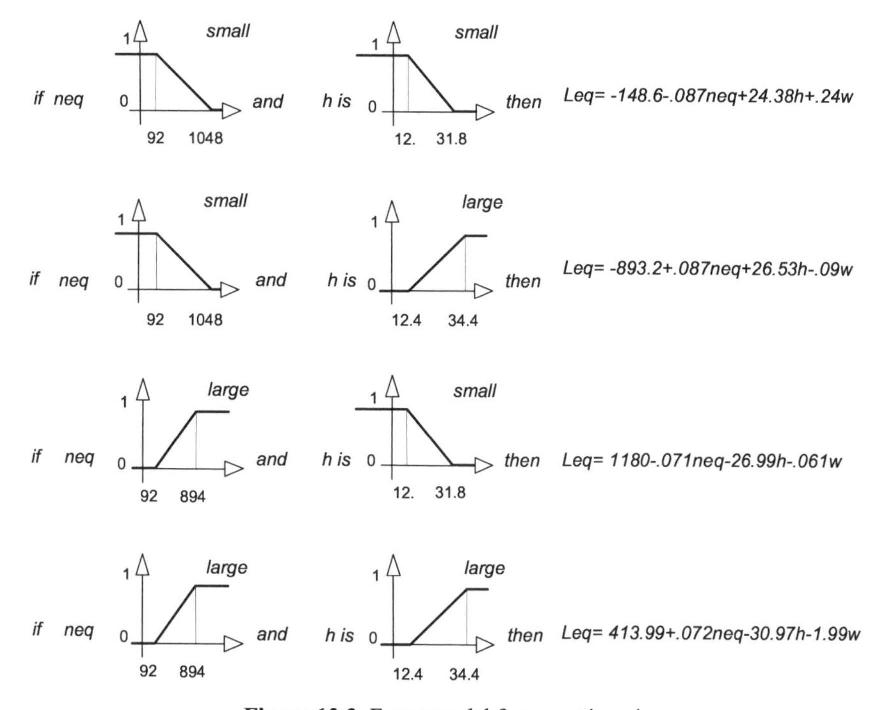

Figure 12.3. Fuzzy model for acoustic noise

The final model obtained is model M220. In fact, the models with eight fuzzy sets were rejected due to the high number of parameters involved.

With respect to the number of measurements of L_{AeqT} (*ca* 350), this condition in fact makes the problem an *ill-posed* one. The model chosen is made up of two fuzzy sets, called *small* and *large* according to the equivalent number of vehicles (*neq*) and the average height of the buildings (h). No fuzzy set was considered for road width.

However, the last mentioned variable was considered in the consequents of the fuzzy rules. The fuzzy model obtained contains four rules and is reported in Figure 12.3.

Figure 12.3 reports the values of those parameters which identify the fuzzy sets of the premises and the coefficients of the consequents.

The performance of the model proposed was tested using both the data employed for estimating the model and the data not taken into account when considering its identification. The results obtained with the above-described fuzzy model for the set of test data were compared both with those obtained with the model represented in Equation 12.5 and with the measured data. This comparison is reported in Figure 12.4.

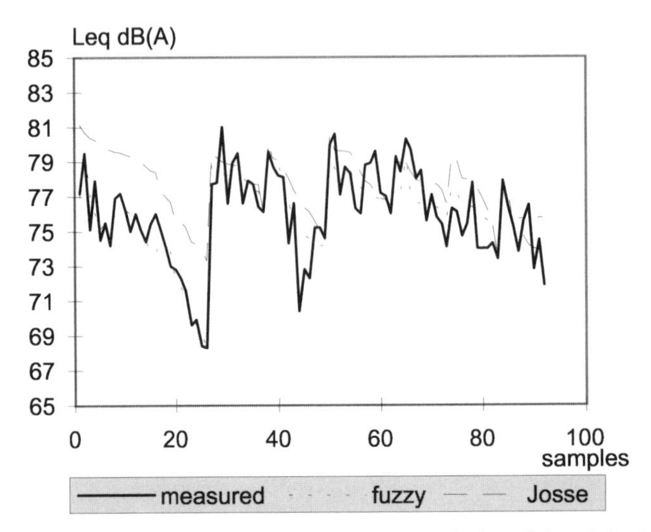

Figure 12.4. Comparison of the measured data, the classical model, and the fuzzy model

A glance at Figure 12.4 proves that the fuzzy model has a better capacity to interpolate the measured data, compared with the empirical-type classical one. Even if the fuzzy approach displays good performance, the optimization phase needed for obtaining the model requires considerable computational effort. To avoid these drawbacks, an approach based on the use of neural networks, and outlined in the following section, was considered.

12.5 The Neural Model

On the basis of the capacity of approximating afforded by the multilayer perceptron (MLP), with sigmoidal activation function for the hidden layers [6-7] a *neural approach* was also used for identifying the acoustic pollution model.

The MLP network was used for determining the functional relationship between the level of equivalent sound pressure, the equivalent number of vehicles, the height of the buildings, and the road width, as was done previously for the fuzzy approach.

Therefore, the MLP has three inputs, *i.e.*, *neq*, *h*, and *w*, and a single output represented by the value *Leq*. Among all the above-described measured sets, a subset of 250 data was used for training the MLP with the back-propagation algorithm, whereas a set of 92 data, not used during the learning phase, was left available for validating the model.

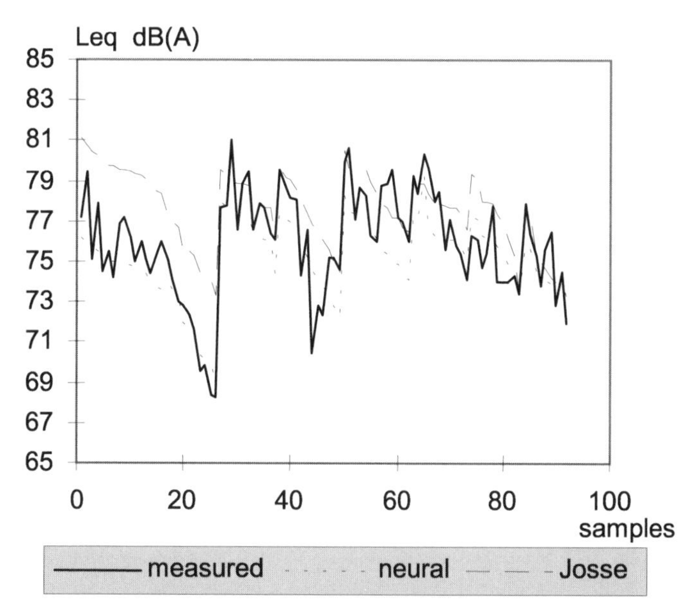

Figure 12.5. Comparison of the classical model, the neural approach, and the measurements

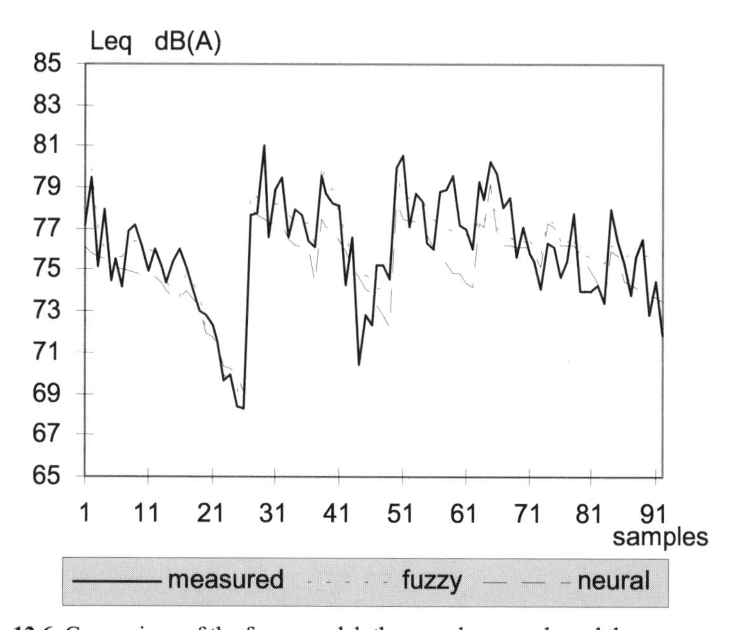

Figure 12.6. Comparison of the fuzzy model, the neural approach, and the measurements

The performances obtained with the neural model are reported in Figure 12.5.

A comparison between the performances of the neural and fuzzy models is reported in Figure 12.6. As can be seen from Figure 12.6, the fuzzy model works better than the neural one, even if the calculation complexity needed for determining a correct fuzzy model is greater. That pointed to the use of the neuro-fuzzy approach with a view to combining the advantages of both these strategies.

12.6 The Neuro-fuzzy Model

With the aim of realizing a *neuro-fuzzy network* model, the structure with a Gaussian-type activation function described in Chapter 9 was adopted [9], [15], [18]. During the learning phase, a set of 250 data was used. In order to determine the FNN topology, two fuzzy sets were chosen for every input variable, thereby obtaining a set of eight rules. With reference to Figure 9.3, the following number of neurons was fixed:

- input level: 3 inputs;
- level A: 6 neurons;
- level B: 6 neurons;
- level C: 8 neurons;
- level D: 1 neuron.

The learning phase requires *ca* 300 iterations. The membership functions obtained for each variable are reported in Figure12.7a, b and c, where *Input 1* is the equivalent number of vehicles, *Input 2* is the height of the buildings, and *Input 3* is the road width. The values of the consequents are reported in Figure 12.7d.

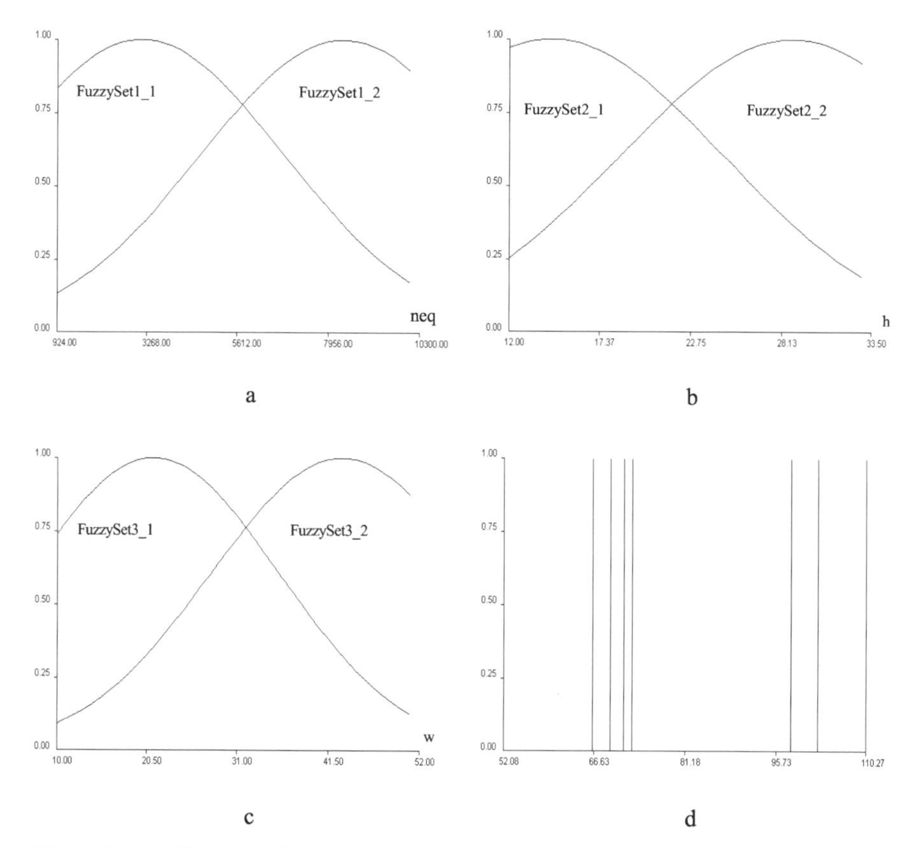

Figure 12.7. a. fuzzy sets for Input 1; b. fuzzy sets for Input 2; c. fuzzy sets for Input 3; d. consequents of the sets of rules

The model obtained has 8 rules as described in greater detail in Tables 12.2, 12.3 and 12.4.

Table 12.2. Fuzzy sets of the input variable

Input Variable Names	Fuzzy Set Names	Centroids
Input Variable 1	FUZZYSET1_1	-0.525801
	FUZZYSET1_2	0.583036
Input Variable 2	FUZZYSET2_1	-0.766005
	FUZZYSET2_2	0.558134
Input Variable 3	FUZZYSET3_1	-0.460455
	FUZZYSET3_2	0.565475

Table 12.3. Fuzzy model

R_1	IF neq is FUZZYSET1_1 AND h is FUZZYSET2_1 AND w is FUZZYSET3_1 THEN Leq is CONS1_1
R_2	IF neq is FUZZYSET1_1 AND h is FUZZYSET2_1 AND w is FUZZYSET3_2 THEN Leq is CONS1_2
R_3	IF neq is FUZZYSET1_1 AND h is FUZZYSET2_2 AND w is FUZZYSET3_1 THEN Leq is CONS1_3
R_4	IF neq is FUZZYSET1_1 AND h is FUZZYSET2_2 AND w is FUZZYSET3_2 THEN Leq is CONS1_4
R_5	IF neq is FUZZYSET1_2 AND h is FUZZYSET2_1 AND w is FUZZYSET3_1 THEN Leq is CONS1_5
R_6	IF neq is FUZZYSET1_2 AND h is FUZZYSET2_1 AND w is FUZZYSET3_2 THEN Leq is CONS1_6
R_7	IF neq is FUZZYSET1_2 AND h is FUZZYSET2_2 AND w is FUZZYSET3_1 THEN Leq is CONS1_7
R_8	IF neq is FUZZYSET1_2 AND h is FUZZYSET2_2 AND w is FUZZYSET3_2 THEN Leq is CONS1_8

Table 12.4. Values of the consequents for each rule

Nomi of the consequents	Values
CONS1_1	0.248745
CONS1_2	0.777778
CONS1_3	0.508815
CONS1_4	-0.582857
CONS1_5	-0.171717
CONS1_6	0.350870
CONS1_7	0.212121
CONS1_8	-0.030303

Figure 12.8 reports the comparison of the neuro-fuzzy model obtained, the Josse model, and the measured data. Instead, Figure 12.9 shows a comparison between all the models.

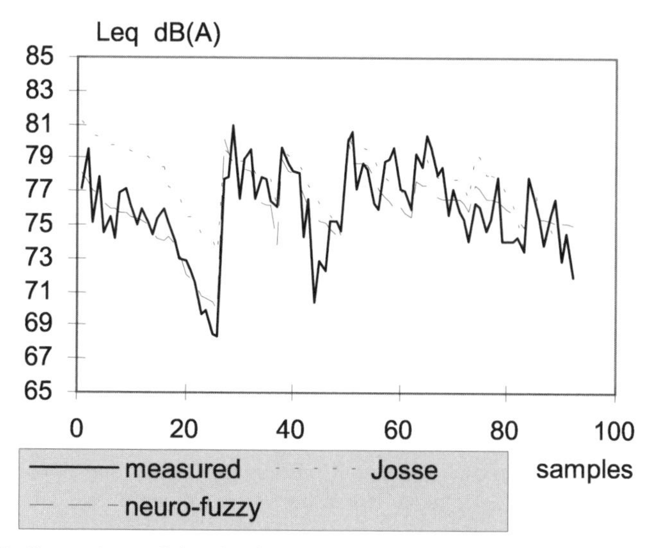

Figure 12.8. Comparison of the classical model, the neuro-fuzzy model, and the measured data

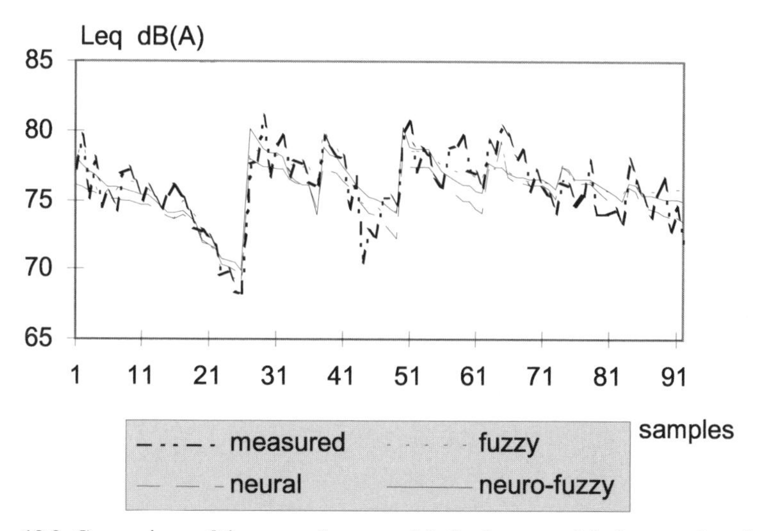

Figure 12.9. Comparison of the neuro-fuzzy model, the fuzzy model, the neural model, and the measured data

As can be observed from the last figure, the neuro-fuzzy model has an intermediate performance with respect to the fuzzy model and the neural one. Given the low computational load of the neuro-fuzzy modeling strategy, compared with the traditional fuzzy approach, its performance may be considered relatively satisfactory.

12.7 References

1. E.P.A. Public Health and Welfare Criteria for Noise. rep. N. 550/9-73-002, 1974
2. E.P.A. Information on Levels of Environmental Noise Requisite to Protect the Public Health and Welfare with an Adequate Margin of Safety. rep. N. 550/9-74-004, 1974.
3. Josse R. Notions d'Acoustique. Ed. Eyrolles Paris, 1972.
4. Lamure C, Auzou DS. Le Niveaux de Bruit au Voisinage des Autoroutes Dégagées. Cahiers du Centre Scientifique et Tecnique du Batiment, Dec. 1984.
5. Burgess MA. Noise Prediction for Urban Traffic Condition. Related to Measurements in the Sydney Metropolitan Area. Appl. Acoust. 1977; 10: 1
6. Don CG, Rees IG. Road Traffic Sound Level Distributions. Journ. of Sound Vib. 1985; 100(1): 41-53
7. Zadeh L. Outline of a New Approach to the Analysis of Complex Systems and Decision Processes. IEEE Trans. on Syst. Man and Cyb. 1973; 3: 1
8. Rumelhart DE, Hinton GE, Williams RJ. Learning Internal Representation by Error Propagation. Parallel Distributed Processing: Exploration in the Microstructure of Cognition. Rumelhart DE, Mc Clelland JL (editor) MIT Press, Cambridge, 1988; 318-362
9. Shing J, Jang R. ANFIS: Adaptative_Network_Based Fuzzy Inference System. IEEE Trans. on Sistem Man and Cybernetics. 1993; 23: 3
10. DPCM 1.3.1991, All. B. "Limiti Massimi di Esposizione al Rumore Negli Ambienti Abitativi e nell'Ambiente Estremo".
11. Beranek L, Ver IL. Noise and Vibration Control Engineering, Wiley Interscience, 1992
12. Harris CM. Handbook of Noise Control. McGraw Hill, 1979
13. Cammarata G, Fichera A, Graziani S, Marletta L. Fuzzy Logic for Urban Traffic Noise Prediction. on J. Acoustic. Soc. Am. 1995; 98; 5: 2607-2612
14. Cammarata G., Cavalieri S., A. Fichera "A Neural Network Architecture for Noise Identification", in Neural Network, vol IV. 1995; 8: (6) 963-973
15. Baglio S, Fortuna L, Xibilia MG, Zuccarini P. Neuro-Fuzzy to Predict Urban Traffic. Proc. EUFIT94, Aachen, Germany, September 1994
16. Takagi, Sugeno M. Fuzzy Identification of Systems and Its Applications to Modelling and Control. IEEE Trans. on Systems, Man and Cybernetics. 1985; 15: 1
17. Press WH, Flannery BI, Teukolsky SA, Vetterling WT. Numerical Recipes: The Art of Scientific Computing, Cambridge, University Press, 1986
18. Horikawa S, Furuhushi T, Uchikawa Y. On Fuzzy Modelling Using Fuzzy Neural Networks with the Back-propagation Algorithm. IEEE Transaction on Neural Networks. 1992; 3: 5

13. Modeling and Control of a Robot for Picking Citrus Fruit

13.1 Introduction

Agriculture is becoming a more interesting sector for application of new automation technologies since improvement of production quality is possible and, at the same time, reduction of costs as well.

From several quarters, groups of researchers have announced the realization of robots for fruit picking and harvesting of vegetables. The aim of the present chapter is to describe a neural network application aimed at the modeling and control of a *mechanical robot* designed for picking citrus fruit [8]. The development of this system was promoted by the Consorzio per la Ricerca in Agricoltura nel Mezzogiorno (CRAM, the Agricultural Research Consortium for Southern Italy), in whose laboratories a prototype, which will be briefly described below, has been built.

The employment of neural networks, and thus of a non-traditional kind of approach, for control application was dictated by the complexity of the robot analytical model, which would have led to difficult control algorithms from the computational standpoint, and by low-level efficiency in performance regarding the specific dynamics of the machine.

13.2 Description of the Robot

The *mechanical structure* of the current version of the robot (the second since the project started) is represented in Figure 13.1. It is articulated in two subsystems: the first, called the *master*, allows the positioning of the other subsystem, called the *slave*, in a certain zone of a *ca* 50m^3 work space. The base of the master subsystem is mounted on a tracked vehicle so that the robot can reach up to the point of fruit picking. The slave subsystem is made up of a *shoulder* on which are mounted two independent *arms*, each equipped with a *viewing device*, by means of which the position of the objective to be reached is identified. The master system is made up of a *kinematic chain* formed by a rotating base to which a four-armed chain is attached by means of rotating prismatic joints driven by hydraulic activators. Each of the two arms of the slave subsystem forms a spherical-type kinematic chain realized by utilizing two rotating-type joints and one prismatic joint.

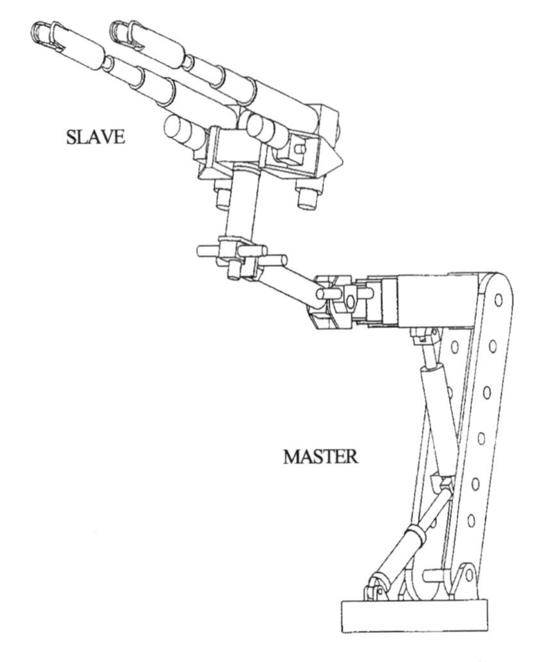

Figure 13.1. Mechanical structure of the robot

The three joints are driven by direct current electric motors. The two rotating joints are connected to the motors by a harmonic drive, while the prismatic joint of the telescopic arm is activated by means of an appropriate belt drive. Below, we will refer exclusively to the functioning of the slave system, since during the picking phase the master system is assumed to be fixed. Furthermore, since the two slave system arms are identical, we will refer to one of them only.

13.3 The Pre-existent Control System

The *basic control system* is represented in Figure 13.2. The reference trajectories are generated by a calculator that processes the acquired images with suitable television cameras mounted at the end of the telescopic arms, supplying the position of the centroid of the object that has to be reached. The three joints of each arm are controlled in uncoupled form by three distinct feedback loops.

The controllers are implemented by axial digital control cards that perform a transfer function of the type:

$$G(z) = k\frac{z-a}{z-b} + \frac{k_i}{z-1} \tag{13.1}$$

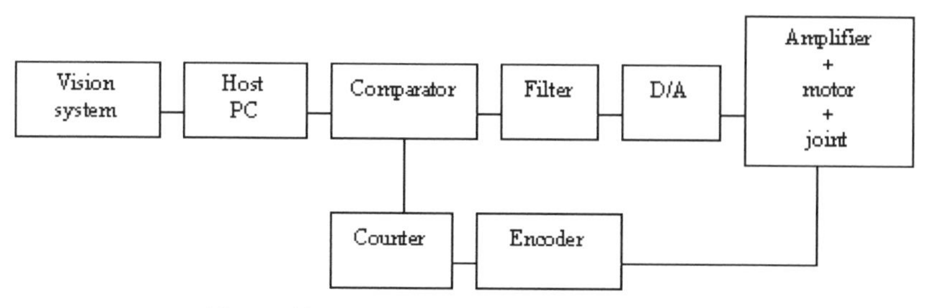

Figure 13.2. Control diagram of a single joint

13.4 Dynamic Modeling of the System

The *dynamic* model of the described system was obtained by employing the Lagrange-Eulero formulation. The references were fixed on the robot following the Denavit-Hartenberg notation. Particular attention was addressed to modeling the telescopic arm, in that it was composed of three elements destined to achieve only one degree of freedom. This study led to the following equations that describe the dynamics of the three joints:

Rotating joints:

$$\begin{aligned}\tau_1 = &\left[aq_3^{\,2} + bq_3 + c + c1 + J_{mot}\right]\ddot{q}_1 + \left[2aq_3 + b\right]\cos^2(q_2)\dot{q}_1\dot{q}_3 \\ &-\left[2\left(aq_3^2 + bq_3 + c\right)\cos q_2 \sin q_2\right]\dot{q}_1\dot{q}_2 \\ &-\left(dq_3 + e\right)\sin q_1 \cos q_2 + F_{a1}\dot{q}_1;\end{aligned} \tag{13.2}$$

$$\begin{aligned}\tau_2 = &\left[aq_3 + bq_3 + c + J_{mot}\right]\ddot{q}_2 + \left[2aq_3 + b\right]\dot{q}_2\dot{q}_3 \\ &+\left[\left(aq_3^2 + bq_3 + c\right)\cos q_2 \sin q_2\right]\dot{q}_1^2 \\ &-\left(dq_3 + e\right)\sin q_2 \cos q_1 + F_{a2}\dot{q}_2\end{aligned} \tag{13.3}$$

Prismatic joint:

$$\begin{aligned}\tau_3 = &\left(m_{eq} + m_{mot}\right)\ddot{q}_3 + \left(hq_3 + g\right)\cos^2(q_2)\dot{q}_1^2 \\ &+\left(hq_3 + g\right)\dot{q}_2^2 + p\cos(q_1)\cos(q_2) + F_{a3}\dot{q}_3\end{aligned} \tag{13.4}$$

where: q_i (i = 1,2,3) are the three co-ordinates of the joints, 1, 2, and 3 are the generalized forces applied to the individual joints, and the remaining elements are parameters of the robot that depend on the physical and geometric features of the system. These equations were obtained with the help of a symbolic calculus code.

Some of the parameters in question could have been obtained from the physical characteristics of the robot. However, that would have meant a considerable loss of time. It was preferred to make use of the availability of the prototype to perform experimental tests for identifying the unknown parameters. In particular, a simulator of the robot's dynamics was implemented, on the basis of the equations reported above.

Some specific tests were designed for the robot applying torque steps to each joint and measuring the corresponding positions by means of encoders.

The data thus obtained were used as the target in the simulation program, allowing us, by means of an iterative process, to obtain the optimal parameter values for the model.

By way of example, Figure 13.3 reports the computed and measured trends regarding the spherical joint azimuthal rotation of three different elongation values for the telescopic arm.

The test in question had the purpose of identifying the inertial term coefficients of Equation 13.3 and the friction term F^{a_1}.

The values reported in Figure 13.1 subsequently allowed us to obtain by interpolation the estimates for the parameters *a, b,* and *c.*

The model identified is the following:

Rotation joints:

$$\begin{aligned}\tau_1 = & \left[1.25q_3^2 - 0.5q_3 + 6.7\right] * q_1 + \left[2.5q_3 - 0.5\right]\cos^2(q_2)\dot{q}_1\dot{q}_3 + 50\dot{q}_1 \\ & - \left[\left(2.5q_3^2 - q_3 + 2.88\right)\cos(q_2)\sin(q_2)\right]\dot{q}_1\dot{q}_2 \\ & - \left(32.66q_3 - 16.33\right)\sin(q_2)\cos(q_2)\end{aligned} \tag{13.5}$$

$$\begin{aligned}\tau_2 = & \left[1.25q_3^2 - 0.5q_3 + 6.7\right] * q_2 + \left[2.5q_3 - 0.5\right]\dot{q}_2\dot{q}_3 + 50\dot{q}_2 \\ & - \left[\left(0.25q_3^2 - 0.5q_3 + 1.4\right)\cos(q_2)\sin(q_2)\right]\dot{q}_1^2 \\ & - \left(32.66q_3 - 16.33\right)\sin(q_2)\cos(q_1)\end{aligned} \tag{13.6}$$

Prismatic joint:

$$\begin{aligned}\tau_3 = & 18\ddot{q}_3 + \left(-50q_3 + 20\right)\cos^2(q_2)\dot{q}_1^2 + \left(-50q_3 + 20\right)\dot{q}_2^2 \\ & + 170\cos(q_1)\cos(q_2) + 90\dot{q}_3\end{aligned} \tag{13.7}$$

The model identified served amongst other things for calculating the maximum values of the individual terms contained in Equations 13.3, 13.4, and 13.5. These values are summarized in Table 13.2.

The model obtained was further subjected to validation by comparing it with measurements made on the robot prototype and not utilized in the parameter identification phase.

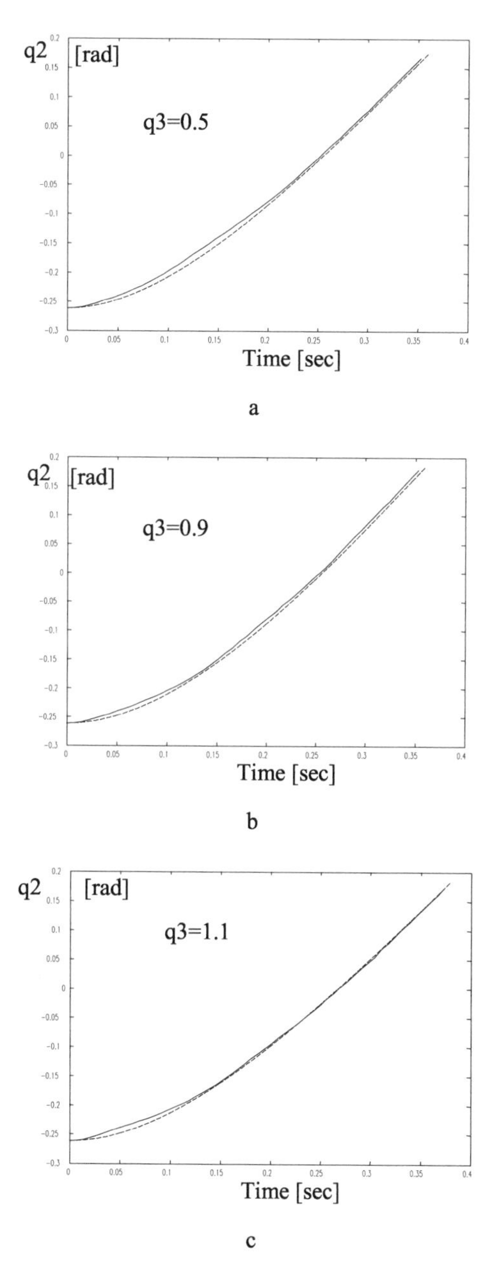

Figure 13.3. a., b. and c. real (—) and simulated (- - -) azimuthal rotation trend for three different telescopic arm elongation values

Table 13.1. Results relative to Figure 13.3

Q3 [m]	$aq_3^2 + bq_3 + c + J_{mot}$	F_{a_2}
Q3=0.5	5.2	38
Q3=0.9	5.7	36
Q3=1.1	6.1	38

Table 13.2. Maximum values for the dynamic equation terms of the robot

q_1, q_2	2rad/sec
$\dot{q}_3$	2m/sec
$\ddot{q}_1, \ddot{q}_2$	5rad/sec^2
$\ddot{q}_3$	8m/sec^2
$\left[aq_3^2 + bq_3 + c + c_1 + J_{mot}\right]\ddot{q}_1$	40Nm
$\left[2aq_3 + b\right]\cos^2(q_2)\dot{q}_1\,\dot{q}_3$	9Nm
$\left[2\left(aq_3^2 + bq_3 + c\right)\cos(q_2)\sin(q_2)\right]\dot{q}_1\,\dot{q}_2$	9Nm
$\left(dq_3 + e\right)\sin(q_1)\cos(q_2)$	20Nm
$\left[aq_3^2 + bq_3 + c + J_{mot}\right]\ddot{q}_2$	30Nm
$\left[2aq_3 + b\right]\dot{q}_2\,\dot{q}_3$	9Nm
$\left[\left(aq_3^2 + bq_3 + c\right)\cos(q_2)\sin(q_2)\right]\dot{q}_1^2$	4.8Nm
$\left(dq_3 + e\right)\sin(q_2)\cos(q_1)$	3.27Nm
$\left(m_{eq} + m_{mot}\right)\ddot{q}_3$	150N
$\left(hq_3 + g\right)\dot{q}_1^2$	140N
$p\cos(q_1)\cos(q_2)$	110N

With regard to this, Figure 13.4 reports the computed and measured trend of the joint variables in the case of simultaneous movement of the three joints, produced by applying three joint steps as inputs.

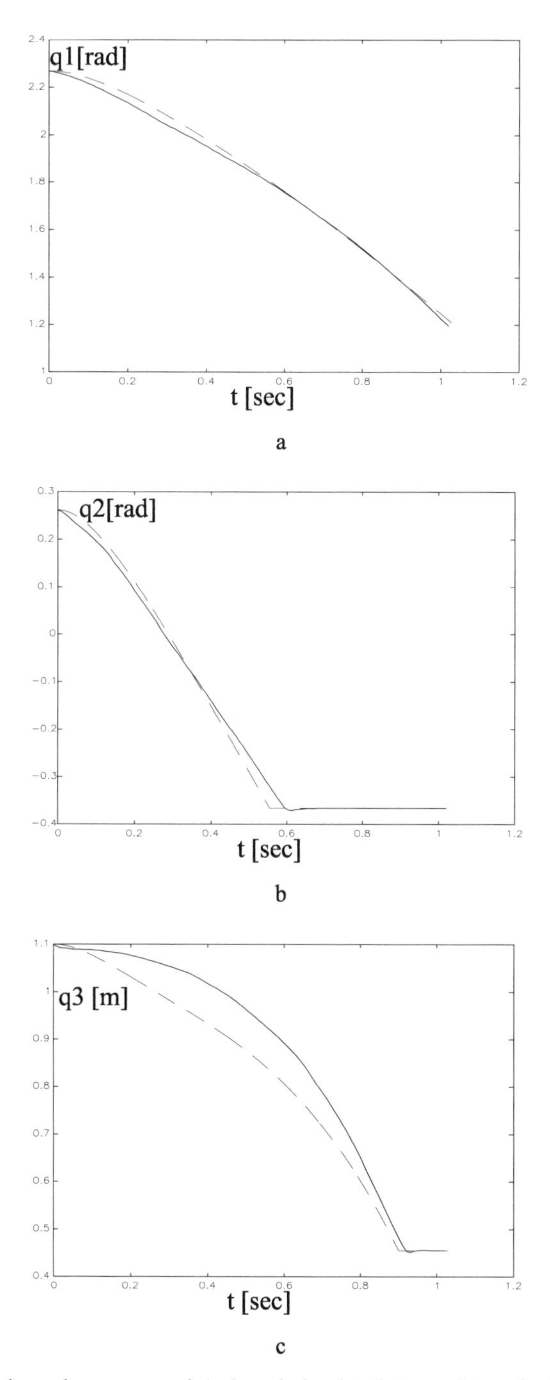

Figure 13.4. a., b. and c. measured (—) and simulated (- - - -) trend of a simultaneous movement of the three joints

13.5 Controlling the Robot

One of the objectives we intended to pursue by formulating the robot model described above was to design a new *control system* instead of the preexistent one, in order to improve the system performance. In fact, the availability of an analytic model, despite its being affected by uncertainties (approximate knowledge of parameters, presence of non-modeled dynamics, friction, *etc.*), allows us to jettison the usual strategy of independent control for the individual joints, based on the use of traditional axial control cards of the type mentioned above.

In fact, regarding robot control, the literature is extremely rich in multijoint control methods, ranging from the traditional computed torque technique to variable structure control, and to adaptive control, *etc.* [1], [5].

Pursuing a research sector which appears particularly promising in the field of robot control, we decided to study a multijoint control scheme based on the use of a neural network. The scheme in question is shown in Figure 13.5.

The role of the neural network is to execute the dynamics inversion, while, in the last analysis, the whole control system effects the traditional feedback linearization method.The theoretical justification for such an approach is reported in [4] and is summarized below:

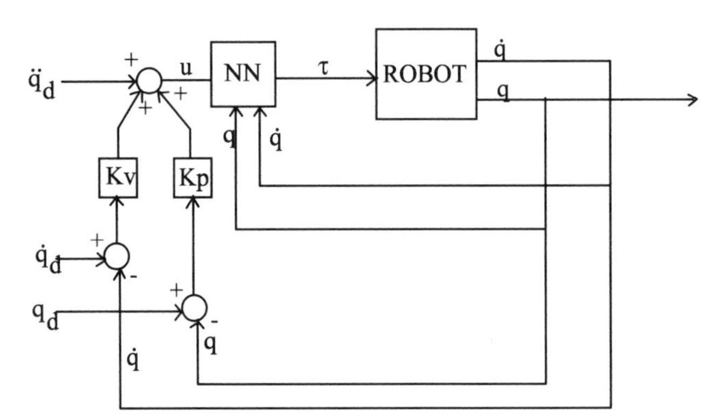

Figure 13.5. Neural multijoint control diagram

With Equations 13.5, 13.6 and 13.7 written in the most compact form:

$$y = D(q)q + h(q,q) + C(q) = F(\ddot{q},\dot{q},q) \tag{13.8}$$

where $q = [q_1 \; q_2 q_3]^T$ in the hypothesis that the neural network exactly interpolates the inverse dynamic function $F(\ddot{q},\dot{q},q)$, the control law $\tau_c = F(u,\dot{q},q)$ where:

$$u = \ddot{q}_d + K_v(\dot{q}_d - \dot{q}) + K_p(q_d - q)$$

ensures that the closed-loop system satisfies the equation:

$$\ddot{e} + K_v \dot{e} + K_p e = 0$$

where $e = q_d - q$.

An appropriate choice of the constants K_v and K_p thus allows us to obtain the desired dynamic behavior of the feedback system. The choice of a neural network in the previously shown diagram, with respect to a traditional model of dynamic inversion, offers many advantages.

First of all, the maturing of experience regarding application of neural networks for identifying non-linear dynamic systems shows the versatility of these schemes with respect to traditional identification procedures. In fact, many researchers by now consider neural tools as the standard easy-to-use ones for identifying non-linear systems [6], a sector in which other approaches will unlikely enjoy analogous consensus of opinion. In addition, the model identification phase by means of a neural network can, in principle, be extended even during the functioning of the system, thereby realizing a typical situation of adaptive control schemes and so making the system capable of facing any variations which might occur in the robot dynamics.

To all this, we should add the prospective of being able in the coming future to use hardware structures devoted to neural systems implementation; these might well prove advantageous, with respect to the traditional architectures, for calculating the inverse dynamics of a robot.

However, we should not minimize the difficulties which may derive from a need to obtain input-output sets of measurements (training sets) that are sufficiently representative of the robot dynamics. Precisely to overcome these difficulties, in the present study, the main purpose of which was to evidence some advantages in terms of neural control robustness compared with the traditional multijoint control schemes, we have made use of the previously described theoretical model for generating neural network training patterns.

From the tests carried out, we found it advisable to face the problem of identifying the robot inverse dynamics dealt with here, by using a multiplayer perceptron having the topology pictured in Figure 13.6.

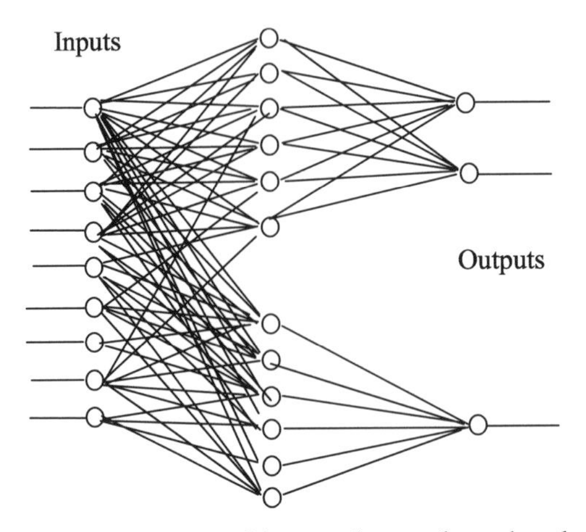

Figure 13.6. Structure of the neural network employed

Notice that the network has nine neurons in the input layer, corresponding to the three joint variables and to their first and second derivatives, and three neurons in the output layer, corresponding to the three torques.

Notice further that in reality the network under consideration is made up of two distinct sub-networks, having only the inputs in common, each one equipped with six neurons in the only hidden layer. Apparently, this unusual topology proved adequate for solving the problems deriving from the different structure of the equations relative to the two rotation joints with respect to the prismatic one and to the different amplitude in ranges of the torques τ_1 e τ_2 with respect to τ_3.The network training patterns were generated, as mentioned above, by exploiting the model and randomly generating the input vector.

The corresponding output values (the torques) were normalized in the interval [0.2 0.8], so as to guarantee compatibility with the network's characteristics, using the formula:

$$\tau_{i\_norm}(k) = \left[\frac{\tau_i(k) - \tau_{i\_MIIN}}{\tau_{i\_MAX} - \tau_{i\_MIN}} \right] * 0.6 + 0.2.$$

$$\text{k} = 1 \ldots 1000; \; i = 1,2,3.$$

The training was done using the learning back propagation algorithm [2] and employing 1000 patterns.

With the aim of validating the proposed neural control strategy, several sinusoidal tracking tests trajectory of different frequency were carried out in simulation, making the individual joints move simultaneously.

The results of these simulations are summarized in Figure 13.7, where by way of comparison we also show the tests corresponding to the use of the preexistent system whose characteristics were illustrated above.

It appears relatively evident that the neural system affords a better tracking capacity. Also the trend of the torques occurring in the two cases at the individual joints should be noted; a greater tendency to saturation of the torque values is found when employing traditional regulators.

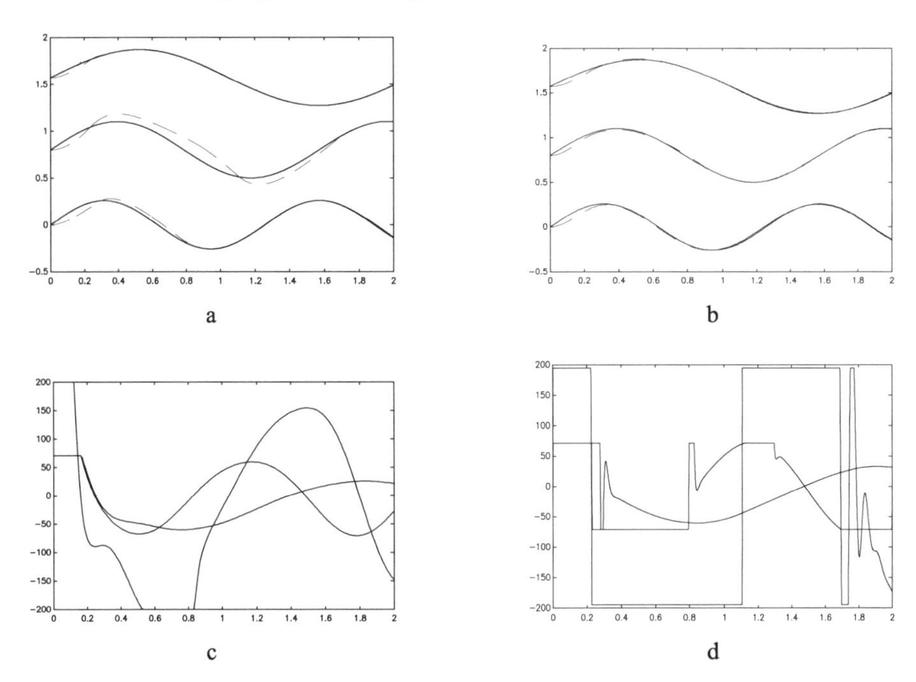

Figure 13.7. a. pre-existent regulator; b. neural regulator; c. pre-existent regulator torque; d. neural regulator torque

13.6 Neural Control with On-line Learning

As mentioned above, one of the advantages of the neural control scheme utilized here is that it can work adaptively according to the scheme shown in Figure 13.8.

The functioning of the control scheme is as follows: with every iteration, the neural network performs a computation of the torques on the basis of q, $\dot{q}$, and u (with the meaning of the symbols being obvious); every N iterations at the vector u, the vector of the measured accelerations is substituted and is thus able to carry out a network learning cycle.

During this phase, the torque output value is not applied directly to the robot, but is compared with the actual torque value, corresponding to the measurements of q, $\dot{q}$ and $\ddot{q}$, for calculating the error to be used in the back-propagation algorithm. Several trajectory tracking simulations were carried out using the above-mentioned adaptive neural control scheme in the presence of perturbation of the robot model.

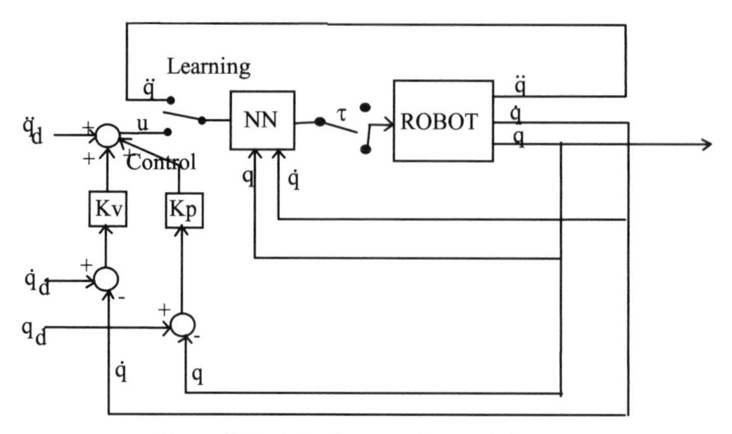

Figure 13.8. Adaptive neural control diagram

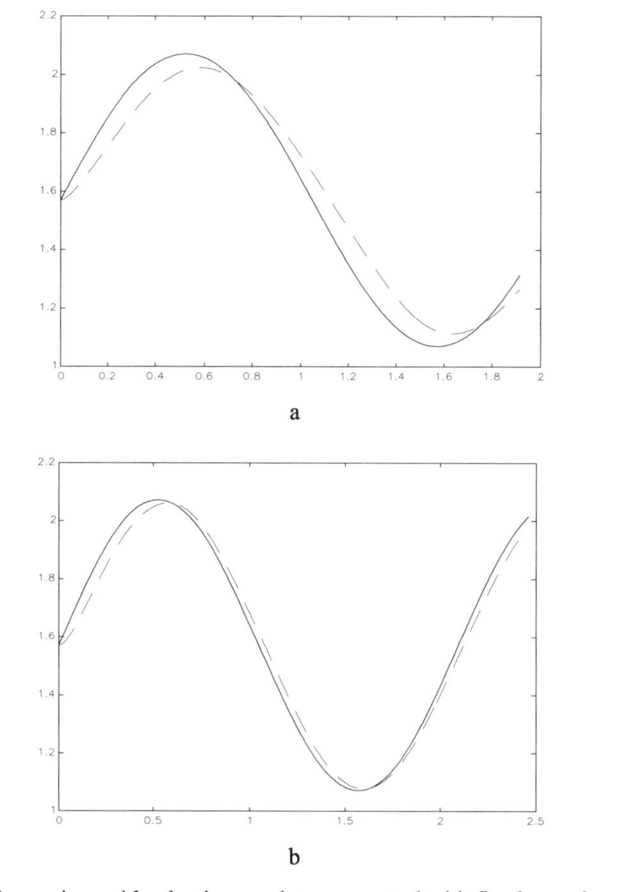

Figure 13.9. Comparison with adaptive regulator: a. control with fixed neural network: Joint 1; b. control with on-line learning network: Joint 1

Similarly, the same tests were carried out using the previously employed fixed neural controller. Generally, these tests evidenced a superiority of the adaptive controller. By way of example, Figure 13.9 reports the comparison between the two controllers. It should, however, be pointed out that in some simulations, corresponding to models with highly perturbed inertia and friction, a tendency toward closed-loop system destabilization was observed.

This behavior is essentially attributed to the fact that the learning process of the new model proceeded slowly with respect to the amount of disturbance applied. However, in all these cases destabilization was also observed in the controlled system with fixed neural network (Figure 13.10).

It can be observed that, in the case of a fixed neural controller, the tracking error is considerable right from the start of the test.

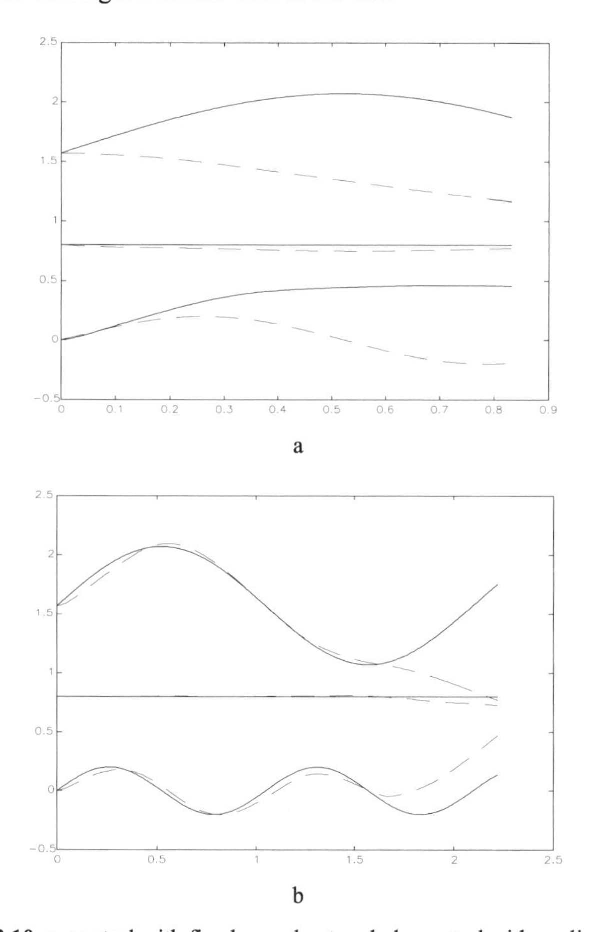

Figure 13.10. a. control with fixed neural network; b. control with on-line learning

13.7 Feed-forward Neural Control

Another scheme, that shown in Figure 13.11, was considered together with the above-reported study.

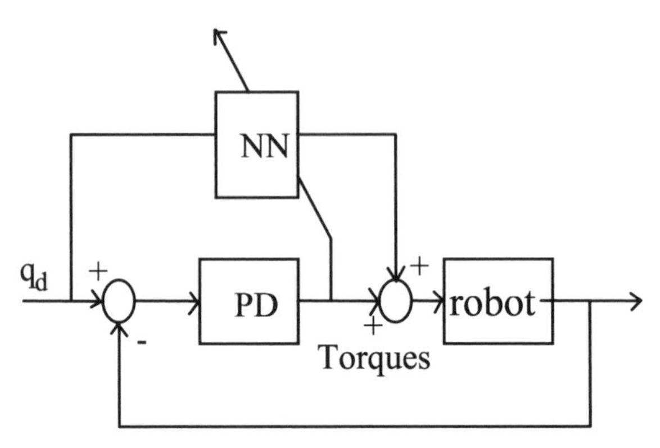

Figure 13.11. Feed-forward neural control diagram

This scheme, too, uses a neural network, but in a different role to the schemes presented previously.

The basic idea in this case is to train the network in such way that it carries out a feed-forward type of control action [3]. The peculiarity of this scheme is that it backs up the traditional control loop without completely substituting it. That allows it to respect preexistent control architecture, as in the case of the robot in the present study, and to improve its performance.

The algorithm consists in making the robot track the same trajectory several times, using at the end of every tracking operation the torques produced by the feedback regulator as the neural network learning target. In this way, the neural network feed-forward action will tend to compensate the errors depending on the non-linearity of the system. The trend of the tracking error on a parabola-line-parabola type of test is shown in Figure 13.12 in three different states of learning for the three joints. In particular, ten simulations were carried out, each one followed by one hundred neural network learning cycles. The simulations carried out showed excellent control scheme behavior, even in the event of a reduced number of learning cycles.

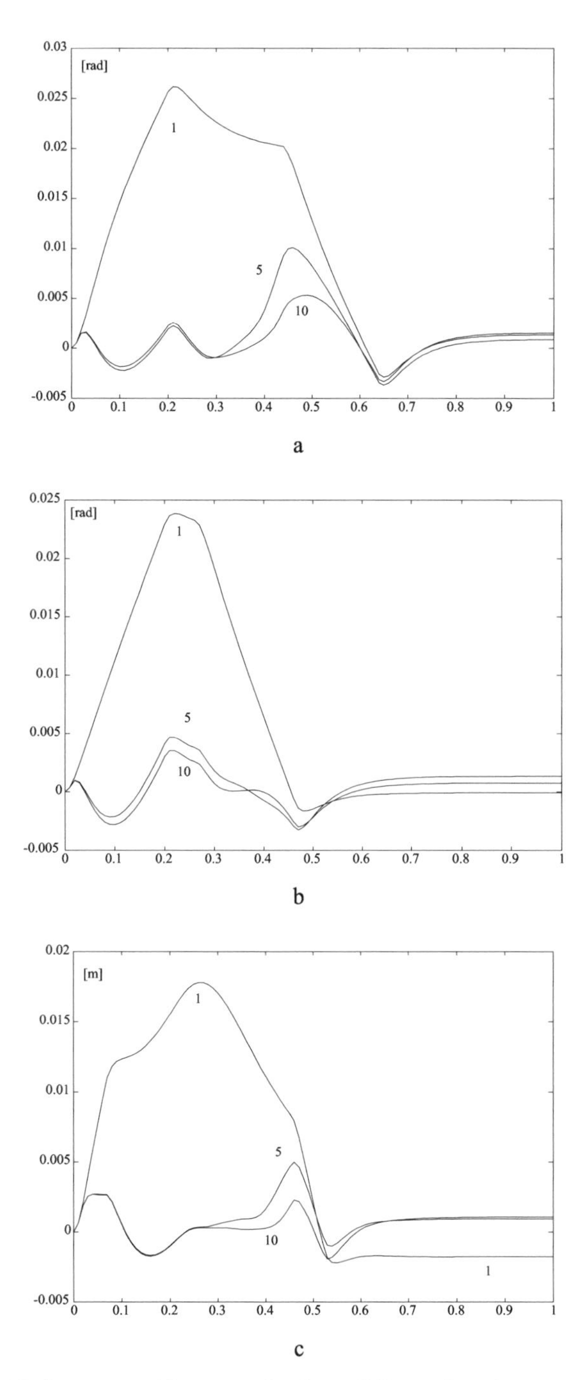

Figure 13.12. Trajectory tracking error in three different learning states. The numbers beside the curves represent the number of learning cycles carried out: a. Link Error 1; b. Link Error 2; c. Link Error 3.

13.8 References

1. Fu KS, Gonzales RC, Lee CSG. Robotics: Control, Sensing, Vision and Intelligence. McGraw-Hill, New York, 1987
2. Mc Clelland J, Rumelhart D, PDP Research Group. Parallel Distributed Processing 1. M.I.T. Press, Cambridge, MA, 1986
3. Newton TR., Yangsheng X. Neural Network Control of a Space Manipulator. In IEEE Control Systems Magazine. 1993; 13: (6): 14-22
4. Poo AN, Ang MH Jr., Teo CL, Qing Li. Performance of a Neuro-model-based Robot Controller: Adaptability and Noise Rejection. In Intelligent Systems Engineering – IEE. 1992; 1: (1): 50-62.
5. Slotine JJE., Li W. On the Adaptive Control of Robot Manipulators. In The International Journal of Robotic Research. 1987; 6: 3
6. Sjöberg J, Hjalmarsson H, Ljung L. Neural Networks in System Identification. In Proceedings of 10th IFAC Symposium on System Identification. Copenaghen. July 1994; 49-71
7. Witcomb LL., Rizzi AA, Koditschek DE. Comparative experiments with a new adaptive controller for robot arms. In IEEE Transactions on Robotics and Automation. 1993; 9: (1): 59-69
8. Fortuna L, Muscato G, Nunnari G, Pandolfo A, Plebe A. Application of Neural Control in Agriculture: An Orange Picking Robot. In Acta Horticulturae. April 1996; 441-450

14. Modeling and Control of RTP Systems

14.1 Introduction

In the case study reported in this chapter, it will be evidenced how soft computing techniques, and in particular the integration of analytical and neural-type procedures, will allow us to model and control the complex systems encountered in the semiconductor industry.

The future of this industry depends very much on the possibility of developing machinery for fabricating wafers which will allow production costs to be reduced and the quality of operation processes to be improved. To lower production costs, ovens must be produced that, in the same chamber, will be able to perform several operations on silicon wafers; in addition, the quality of the individual processing phases must be improved in order to reduce the number of defective wafers. In a such a context, the role of control systems appears a decisive one.

In the field of processing machinery, many hopes have been placed in the so-called rapid thermal processing (RTP)-type ovens which will be able to perform all the basic operations on wafers: firing, anodizing, and chemical vapor deposition. In this apparatus, temperature control is of the utmost importance, as has already been pointed out by various authors [1], [5], [7], in that rapidly variable thermal profiles must be dealt with [4], and at the same time the necessary uniformity of temperature inside the oven must be guaranteed.

14.2 Structure of a RTP Oven

The diagram of a *RTP oven*, of the kind produced in prototype form at Stanford University, is presented in Figure 14.1.

Unlike the traditional semiconductor ovens, in which the operation processes may require hours of time, because of the considerable thermal capacity involved, in a RTP-type oven, only the wafers are heated or cooled while the oven walls are kept at room temperature by a forced air conditioning system. This is the explanation as to why, in a RTP-type oven, heating or cooling of wafers can be carried out in the order of seconds.

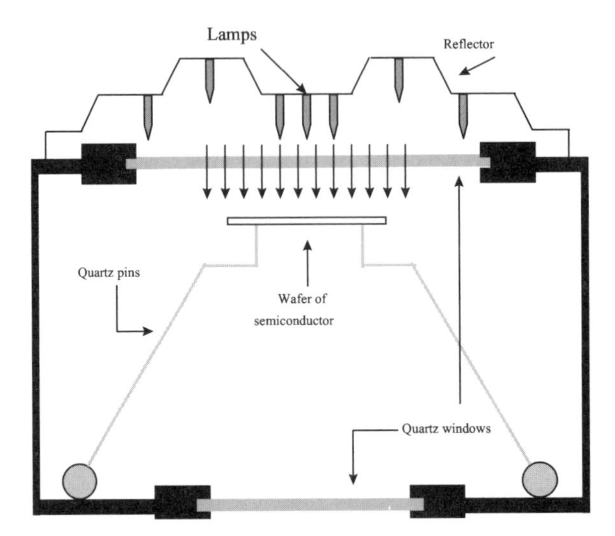

Figure 14.1. Section of a RTP oven

In the RTP oven section in Figure 14.1, it can be seen how the *semiconductor wafers* are heated by *infrared lamps*, usually of halogenous or tungsten type, arranged in three different rings each of which is fed by an independent light source. That allows the control system greater flexibility and greater heating uniformity. Instead, the semiconductor wafer is placed on an appropriate support and is thermally insulated by quartz pins. Essentially, the wafer is heated by irradiation and only on the surface is it exposed to the action of the lamps through an appropriate window.

A more accurate analysis of the heating mechanism in question evidences the additional presence of convective and conductive-type phenomena. The temperature inside the chamber can be measured both by pyrometers and, as in the case in the figure, by thermocouples arranged in the lower quartz window, following the type of geometry illustrated in Figure 14.2.

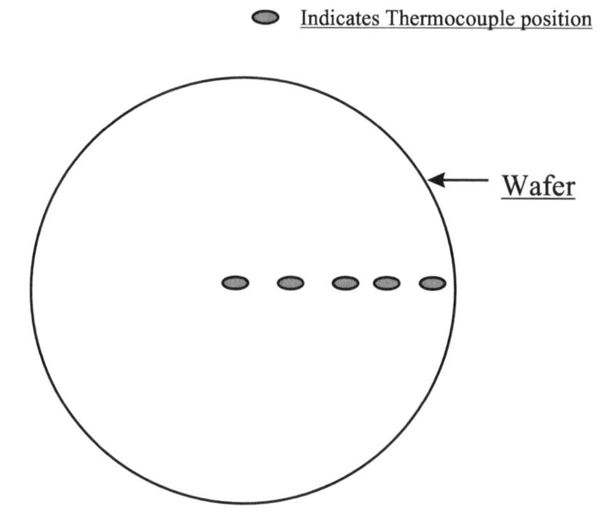

Figure 14.2. Position of temperature sensors

14.3 Model of RTP Oven

The first step to be taken in controlling the type of system described is to develop a model. In fact, under close analysis, the system is of the non-linear (witness the heating mechanisms involved) distributed-parameter type.

At the same time, development of an excessively accurate model would pose a problem for the development of the control system. A compromise between the requirements of accuracy and those of simplicity is to treat the system as one of non-linear concentrated-parameter MIMO-type, *i.e.*, with several inputs (the supply voltage of the lamps) and several outputs (the temperature measured at the points where the temperature sensors are installed). However, even a non-linear model, although having concentrated parameters, may prove difficult to treat with traditional control methodologies.

With respect to this, see the work of Gyugyi *et al.* [4], who, while attempting to develop traditional linear controllers of the linear quadratic Gaussian (LQG) type, identified several linearized models of the system around particular operation points.

Although such a choice may lead to considerable simplification in the model identification phase, it may also have drawbacks both with regard to the overall system performance and to the complications arising from the management of various linear controllers.

In this context, to fall back on neural networks represents a turning point both in the difficulty of identifying a complex non-linear system and in the implementation of the relative control system, as will be shown later in this chapter.

14.4 Identifying a RTP System Neural Model

The data used for identifying the *neural model* were dealt with in a work of Gyugyi *et al.* [3]. In fact, in that work they reported four linear models of the RTP system around the regime operation points at temperatures of 400 °C, 420 °C, 500 °C, 600°C, and 700 °C.

These models were simulated using binary-type input signals with random sequence (pseudo-random binary sequence (PRBS)) of the kind illustrated in Figure 14.3.

By means of these simulations, a sufficient number of patterns can be obtained for training a neural network whose purpose is to model the RTP system. To be precise, in the case in question, patterns generated with the linearized models around temperatures of 500 °C and 700 °C were employed for training the network, with the model linearized around 700 °C for the validation phase.

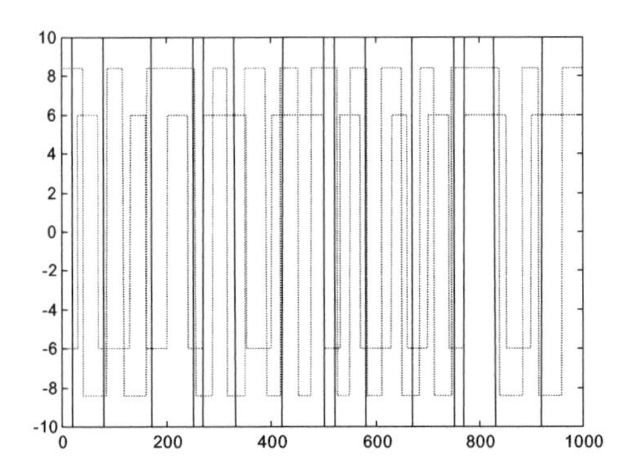

Figure 14.3. Voltage signals used as inputs for modeling

An example of linearized model output around a temperature of 700 °C, simulated with inputs of the kind indicated above, is shown in Figure 14.4.

The structure of the patterns employed for identifying the model is the following:

$$[y_1(k),y_2(k),\dots,y_5(k)]=f(u_1(k),u_2(k),\ u_3(k),\ y_1(k\text{-}1),y2(k\text{-}1),\dots,y5(k\text{-}1)]$$

where:

$y_i(k)$ is the temperature measured at the time k by the i-th thermocouple;
$u_i(k)$ is the voltage supplied by the lamps of the i-th ring at the time k.

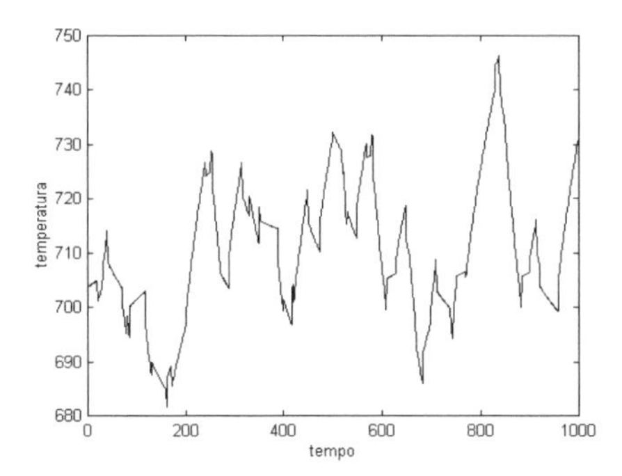

Figure 14.4. Linearized model around a temperature of 700 °C

The neural model structure is given in Figure 14.5.

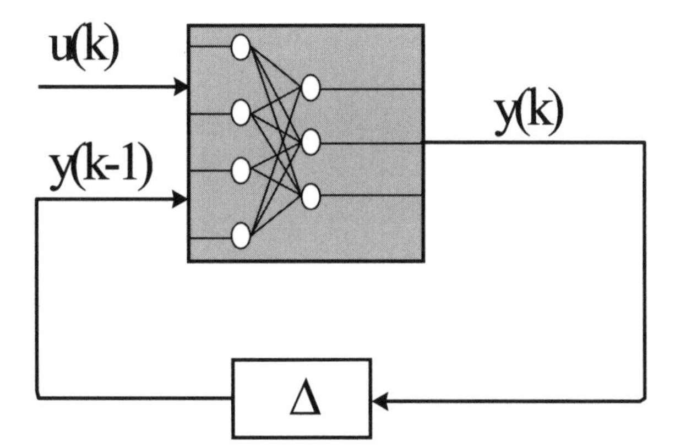

Figure 14.5. Diagram of the neural model considered

After some attempts, it was found convenient to use a multilayer perceptron with Gaussian-type activation functions and an 8-8-5-type topology. The results from the comparison between the identified neural model and that used for generating the training patterns are reported in Figures 14.6-14.10.

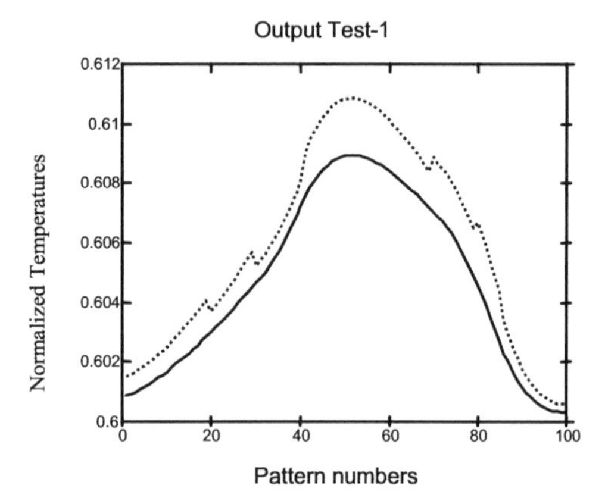

Figure 14.6. Comparison between the neural model output and that of the linear mode

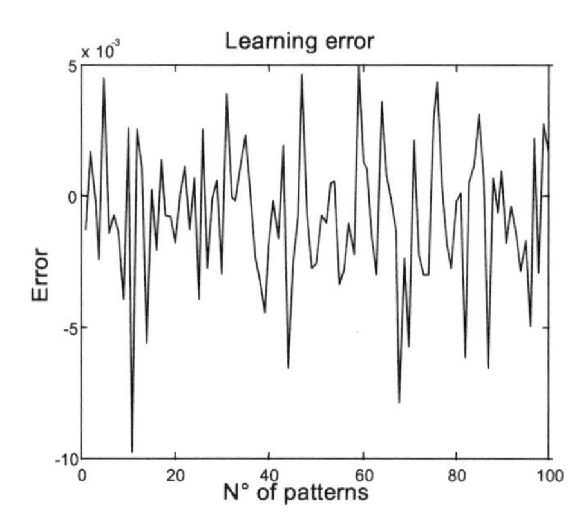

Figure 14.7. Error during the training phase

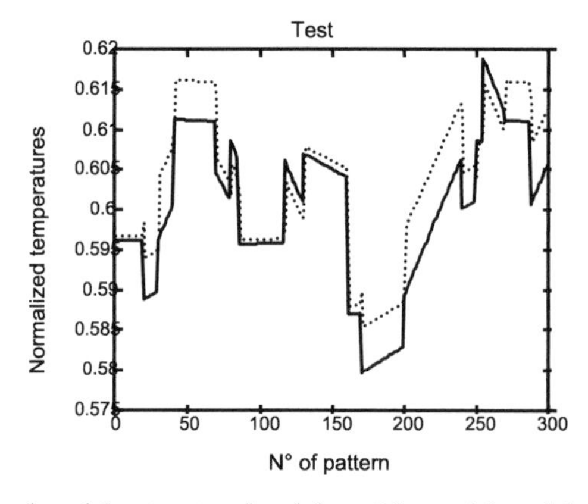

Figure 14.8. Neural model output trend and that of the model used during test patterns

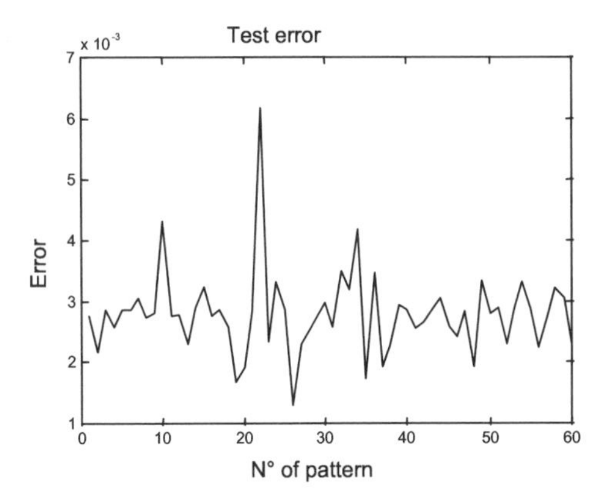

Figure 14.9. Error using test patterns

In order to further validate the results, we considered an additional voltage input obtained by adding a white noise in the range [-1.5 1.5] to a PRBS-type signal, as shown in Figure 14.10. The relative temperature trends for neural and linearized models around an operational point of 700 °C, corresponding to this new input, are reported in Figure 14.11.

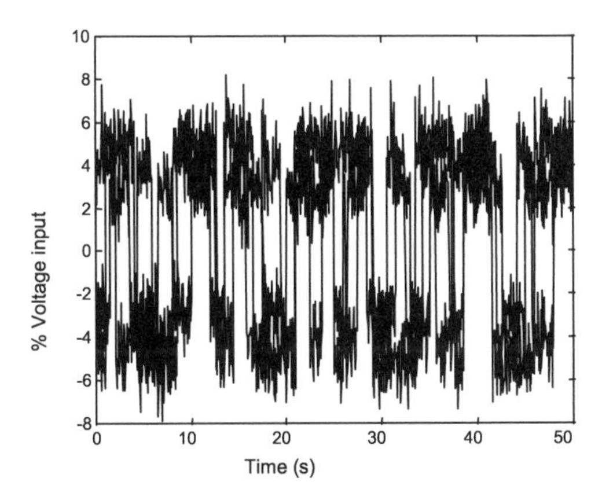

Figure 14.10. Voltage trajectories used for simulation

As can be seen, the two figures practically overlap, and therefore the difference (error) between the two outputs is reported in Figure 14.12 in order to facilitate comparison.

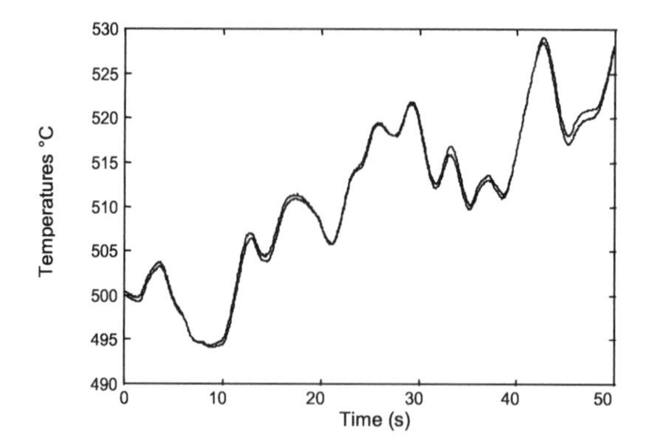

Figure 14.11. Comparison between the neural and linearized models at 700 °C, at the first thermocouple

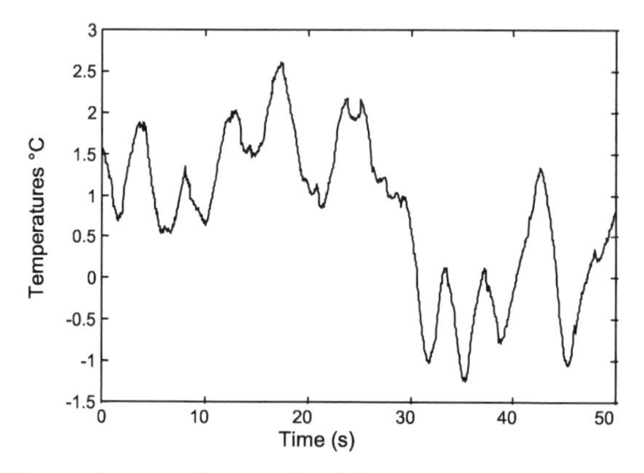

Figure 14.12. Difference between the neural and linearized models (see Figure 14.11)

To demonstrate that the neural model works satisfactorily in the whole temperature range considered interesting, the following figures also show the errors had with linearized models at 500 °C (Figure 14.13), 600 °C (Figure 14.14), and 700 °C (Figure 14.15).

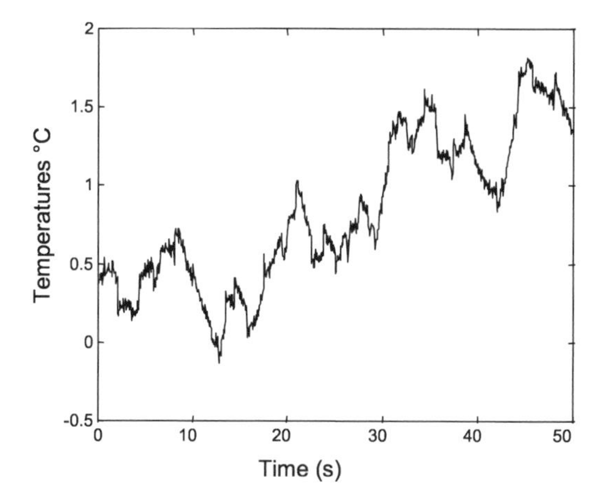

Figure 14.13. Maximum difference between the neural and linear model outputs at 500 °C

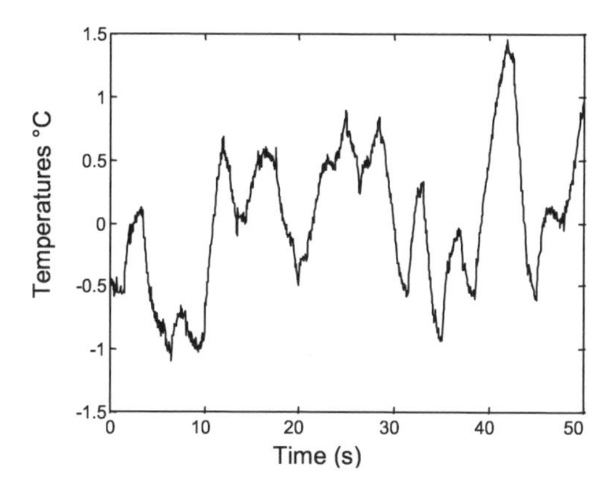

Figure 14.14. Maximum difference between the neural and linear model outputs at 600 °C

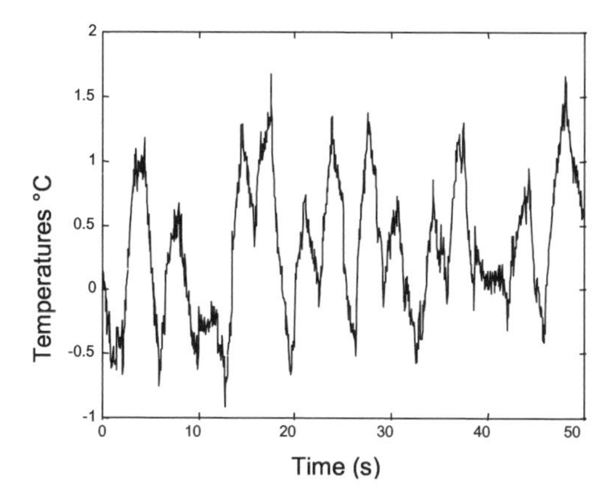

Figure 14.15. Maximum difference between the neural and linear model outputs at 700 °C

It can be understood from the graphs that there is satisfactory agreement between the only neural model identified and the various linearized models used for generating the data required for identification.

14.5 Design of a RTP System Neural Controller

Availability of a RTP system model is a decisive condition, but not the only one, for developing a RTP system which will really be operative. In fact, it is just as important to design an adequate control system. Controlling the temperature in a RTP system, as pointed out in the introduction to the present chapter, requires both that the temperature inside the oven be maintained spatially uniform and the capacity to track rapidly variable thermal profiles. Following the same line of approach as that used for identifying the model, we will describe below a neural controller design able to respond to the control demands mentioned above. This strategy is an alternative to other, traditional, control techniques of the type proposed by Gyugyi [4].

The neural control scheme used is pictured in Figure 14.16.

The controller synthesis strategy is based on the concept of inverse dynamics: in other words, the controller is a particular neural network trained to calculate optimal input voltage for the halogen lamps so as to realize the desired thermal profiles inside the oven. With regard to the diagram of Figure 14.16, the variable y_d is defined as follows:

$$y_d(k) = a \cdot y(k-1) + (1-a) \cdot r(k)$$

where:

- $r(k)$ is the reference thermal trajectory;
- a is a parameter whose value is appropriately chosen to avoid too brusque an input control variation due to too sudden a variation in the reference trajectories.

The neural controller was obtained with an iterative process in which, at every step, a steepest gradient descent type of optimization algorithm calculates the signal control that minimizes a cost function defined as follows:

$$\begin{aligned} J(k) &= [y(k) - y_d(k)]^2 \\ y(k) &= NN(u(k), y(k-1), ..) \\ u(k) &= \min_{u(k)} [J(k)] \end{aligned}$$

where $NN(\bullet)$ stands for the function approximating the system model, obtained, as previously seen, by a neural network.

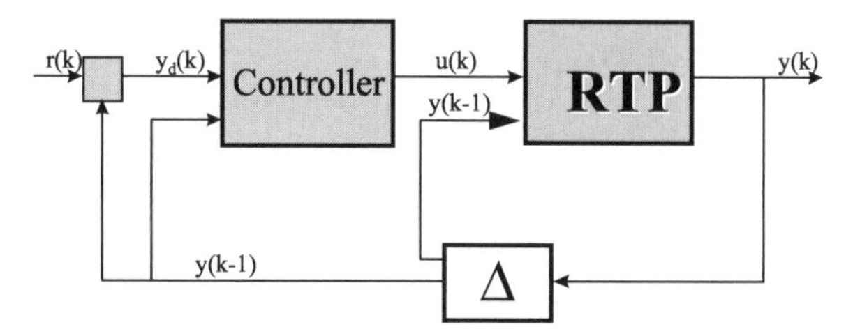

Figure 14.16. Control block diagram

The controller obtained at the end of the optimization process was tested and some of the performances are reported in Figure 14.17. In particular, this figure shows the controlled system output with a step-wise variation in temperature and a regime value of 700 °C starting from an oven temperature of 645 °C. In the simulation, the parameter value a was set equal to 0.4. The various curves reported refer to the temperatures recorded by the five thermocouples visible in Figure 14.2.

The result shown appears all the more significant in view of the fact that the initial condition of the system (645 °C) before the step-wise temperature variation was applied is far from the values of 500 °C, 600 °C, and 700 °C corresponding to the regime values around which Gyugyi *et al.* linearized the models for generating the I/O data in the identification phase of the RTP system neural model.

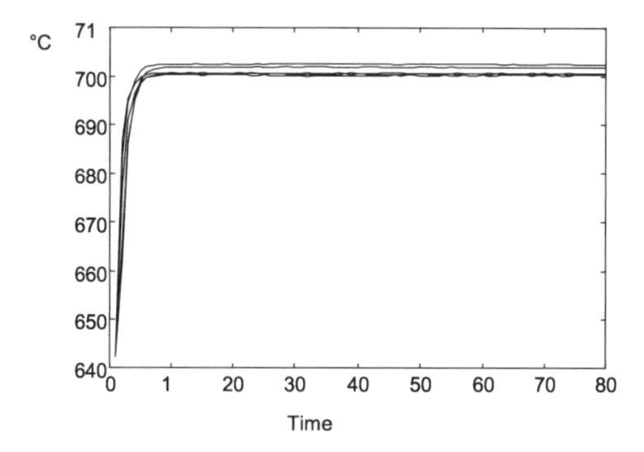

Figure 14.17. Example of the RTP system step response controlled by the neural controller

Lastly, in the following figure (Figure 14.18) the maximum (simulated) difference inside the RTP system is reported with respect to the simulation described above. It can be proved that this difference is contained in a range of ±3°C, which is compatible with the aims of the RTP system.

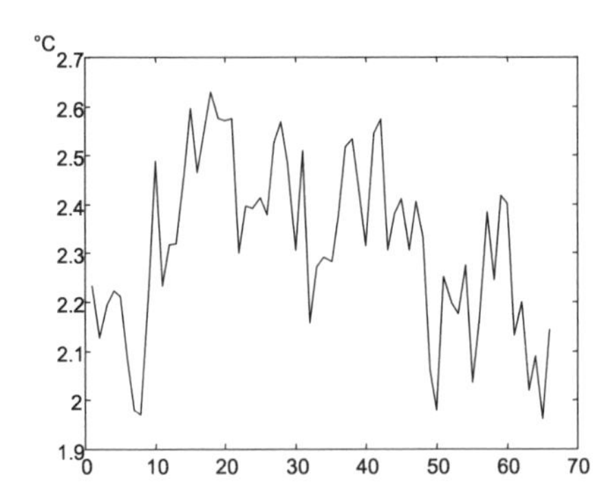

Figure 14.18. Maximum difference in temperature inside the RTP oven chamber for the simulation described in the text

14.6 Conclusions

In this chapter, we have described a neural approach to the problems of modeling and controlling a RTP system. The results reported show the utility and goodness of the approach with respect to other, traditional methodologies. The main advantage is that of being able to identify with back propagation, by now a standard procedure, complex non-linear systems. Thereby we can avoid falling back on the technique of identifying a host of linear models, each one valid in a limited operative sphere of the original non-linear system. These difficulties are even more evident in many cases, in fact like the one described here, in which the purpose of the modeling is control. It is precisely for this last-mentioned aspect that the chapter offers an example of a neural control scheme based on inverse dynamics techniques.

The results reported herein assume even more significant importance when it is observed that the modeling and control procedure described do not take into account the particular dynamic structure of the system and as well, to a certain extent, the number of inputs and outputs that can therefore cover the needs of the majority of applicational problems that we happen to meet in practice.

14.7 References

1. Emani-Naeini A, Kabuli MG, Kosut L. Finite-time Tracking with Actuator Saturation: Application to RTP Temperature Trajectory Following. In Proc. Conf. Decision and Control. December 1994
2. Shaper C, Cho Y, Park P, Normann S, Gyugyi P, Hoffmann G, Balemi S, Boyd S, Franklin G, Kailath T, Saraswat K. Modeling and Control of Rapid Thermal Processing

3. Gyugyi P.J, Man CY, Granklin G, Kailath T, Roy RH. Model-based Control of Rapid Thermal Processing Systems. In The 1st IEEE Conference on Control Applications. Dayton Ohio, 1992; 473-482
4. Gyugyi PJ, Cho YM, Franklin GF, Kailath T. Control of Rapid Thermal Processing: A System Theoretic Approach. in Proc. IFAC World Congress. 1993
5. Campbell SA, Ahn KH, Knutson L, Liu BJH, Leighton JD. Steady State Thermal Uniformity and Gas Flow Patterns in a Rapid Thermal Processing Chamber. IEEE Transaction on Semiconductor Manufacturing. 1991; 4 : (1): 14-20
6. Cybenko G. Approximation by Superposition of a Sigmoidal Function. Mathematics of Control, Signals and Systems. Springer-Verlag, New York, 2: 318-362
7. Gyugyi PJ. Model-Based Control Applied to Rapid Thermal Processing. A dissertation for the degree of Doctor of Philosophy. Stanford University, June 1993
8. Lord HA. Thermal and Stress Analysis of Semiconductor Wafers in a Rapid Thermal Processing Oven. IEEE Transaction on Semiconductor Manufacturing. 1988; 1: (3): 105-114
9. Moor BD, Moonen M, Vandenberghe L, Vandewalle J. A Geometrical Approach for the Identification of State Space Models with Singular Value Decomposition. Proc. IEEE ICASSP. New York, NY, 4: 2244-2247
10. Sjöberg J, Hjalmarsson H, Ljung L. Neural Networks in System Identification. Proceedings of 10th IFAC Symposium on System Identification. Copenaghen. July 1994: 49-71
11. Fortuna L, Muscato G, Nunnari G, Papaleo R. A Neural Networks Approach to Controlling the Temperature on Rapid Thermal Processing. Proc. of 8th Mediterranean Electrotechnical Conference, Bari, May 1996. II: 649-65
12. Schaper CD, Cho YM, Kailath T. Low-order Modeling and Dynamic Characterization of Rapid Thermal Processing. Appl. Phys., A. 1992; 54: 317-326

15. A Neural Network to Predict Air Pollution in Industrial Areas

15.1 Introduction

The atmosphere plays an essential role in absorbing and diffusing chemical substances, both gases and particles. In fact besides its natural gas, such as oxygen, nitrogen, rare gas, steam, it contains, in a certain variable quantity, gases quantities and particles owed both to natural processes and to human activity. In fact several gases, such as sulphur dioxide, hydrogen sulphide and carbon dioxide, are introduced into the atmosphere as products of natural phenomena, such as volcanic activity, vegetation decomposition, wood fires, and so on. Besides *natural pollution*, there exist certain substances that derive from *human activity* which leaves into the atmosphere a great number of wastes, mainly due to carbon, sulphurous, azote and hydrocarbon oxides. Industrial, thermal and thermo-electric plants, besides urban pollution mainly due to vehicles, have heavily affected the atmosphere degradation, giving often rise to great pollution cases. In [1] the effects on earth populations of large-scale perturbations due to pollution are shown. An attempt has also been made in order to investigate about the main damaging elements. The most important gas pollutants are essentially sulphurous dioxide, carbon dioxide, carbon monoxide and azote oxides. Such substances are called *primary pollutants*, because they are responsible of more than the 90% of the whole air pollution. This phenomenon, as it can be argued, is very complex, both for its nature and its effects that can be very different, depending on the topographic position, weather and climate conditions. The importance of the problem at hand as regards its consequences on human and nature life has led to a growing interest of the research world, in order to prevent dangerous environmental situations taking also into consideration that the human polluting activities, both in urban and in industrial areas, rarely decrease with time. Moreover even low concentrations of pollutants can be seriously dangerous.

Recently the possibility of creating monitoring networks for recording pollutant concentrations in the atmosphere has revealed a powerful strategy in order to acquire a considerable knowledge about the dynamics of such phenomena as regards their diffusion and their consequences on human health. Moreover the possibility derived to think about predictive models to control air quality.

The aim of the introduced strategy is to derive *predictive models* of SO_2 pollutant mean value concentration in an industrial area, taking into consideration that its complex dynamics depends on a great quantity of variables, among which the meteorological and climate conditions and the particular geography of the site into consideration. Owing to the complexity of the phenomenon and to the strict dependence on *boundary conditions*, often a classical approach, consisting in the integration of the diffusion equations comes inconvenient. The further problem of the lack of the *source terms*, *i.e.*, the measures of the pollutant at the industrial chimneys makes the use of the classical analytical approach practically unfeasible. The necessity therefore arises to derive a predictive model considering only data drawn from the monitoring stations placed into the industrial area. The neural approach therefore becomes suitable to perform such a task.

15.2 Models of Air Pollution

SO_2 is the most common pollutant: therefore it acts as a guiding index for testing air quality, owing to the presence of sulphurous in almost all combustibles. The complexity of atmospheric phenomena implied serious difficulties in defining *air quality* in its various and real aspects: on the contrary, a law on this subject has to provide simple criteria which could be directly adopted. Under such point of view and after the acquisition of criteria from the World Health Organization and from the European Community, Italy has promulgated a series of lows in 1983 which fix maximum values for the pollutant mean value concentrations and also state the standards for air quality. In particular, for the SO_2 pollutant, the maximum mean value of the concentration allowed for a day considered into the whole year cannot exceed $80\mu g/m^3$.

Taking into consideration, as previously outlined, that classical analytical models cannot be applied and a black-box identification strategy has appeared suitable. In literature several models have been introduced in order to perform black-box estimate of SO_2 pollution levels in industrial areas [2]. In particular hour concentration predictions have been performed. The model employed belong to the class of auto-regressive integrated moving average (ARIMA) models, [3], whose parameters are derived considering, besides pollutant concentration, meteorological data of the place taken into consideration. Moreover a second order stochastic structure has been employed [2]. The results obtained were with no doubt encouraging, but the high non-linearities which characterize such phenomenon suggest the use of neural networks as modeling architectures.

15.3 Air Pollution Modeling via Neural Networks

Multilayer perceptrons (MLPs) are taken into consideration in order to build *predictive models* for SO_2 air pollution in an industrial petrochemical area at the aim to control air quality. Until now data drawn from measures has been used to monitor air quality. Therefore procedures to avoid high pollution situations start only when such situations have already taken place. From such considerations it derives that using available data to build predictive models could avoid many dangerous situations to happen. In order to build such predictive models the use of neural networks is fully justified since such architectures have been established to be universal interpolators, both from the theoretical point of view of continuous functions approximation [4] and for their application to non linear system identification [5-6]. The *industrial area* taken into consideration has twelve centers of data recording, placed in key positions on the area in order to monitor the dynamics of pollutants suitably. The strategy of building a neural network able to estimate the pollutant concentration next to each center has been adopted therefore. By doing this, a network for pollutant estimation over the whole area can be built. Indeed the authors have already been involved in such a problem and they used a unique MLP structure for SO_2 estimation. Though the results obtained were fairly good [7], the testing performance considering patterns built in other seasons did not led to good results. This fact can be justified if we think that the atmosphere is certainly a very complex time-varying system. Therefore, a learning model which *freezes* its *knowledge* in a particular time period, is able to perform the short-term prediction only a narrow time window ahead. From such a consideration the idea has arisen to implement a structure able to *learn on-line*, *i.e.*, a network able to perform a prediction, and later to use the real (measured) value of the pollutant to tune or *adapt* its *knowledge*, represented by its synaptic weight value, to match the new meteorological-pollutant relations. Moreover, further information from the petrochemical production policy has been acquired which has helped to build an efficient architecture. If the mean value of the pollutant concentration in a particular moment goes beyond a certain threshold, a *pre-alarm* situation takes place which obliges the industries to change production suddenly in order not to go beyond the prescriptions imposed by law. Such a fact clearly leads to a certain economic effort. The possibility of having a device able to predict a *pre-alarm* situation could allow the planning of a suitable production policy and so the saving of money. Moreover the prediction of the SO_2 mean value concentration is a very complex task. A way to reduce the complexity associated with each network is to divide the prediction into different steps, each one delivered to a structure able to perform a particular task, thus allowing better performance for the whole architecture.

15.4 The Neural Network Framework

From what previously discussed it can be thought to divide the work of pollutant prediction in a number of simpler tasks: a first *rough prediction*, in which the structure has to *classify* if, on the basis of current meteorological-pollutant data, the predicted pollutant value will reflect a *pre-alarm* situation or otherwise a normal condition. The *refined prediction* will be delivered to two other structures, learned to estimate the crisp value of the pollutant mean value concentration inside one of the two classes which the pattern at hand belongs to. We'll refer to the first step of the prediction (the *rough* prediction) as the *classification step*, and to the following one as the *prediction step*.

All the networks are allowed to learn *sample-by-sample*, both for the classification and for the prediction step, as it will be outlined in the following.

As regards the data taken into consideration, since the SO_2 pollutant dynamics are strictly related to the dynamics of the meteorological parameters, among all the recording stations existing in the industrial area, the station which could offer the richest amount of meteorological information, together with, of course, the measures of the SO_2 pollutant was considered.

Therefore the following quantities have been taken into account for the pollutant prediction: temperature (*T*); relative humidity (*RH*); wind speed (*WS*); wind direction (*WD*); rain level (*R*); solar radiation (*RAD*); pressure (*P*); radio acoustic sound system (*RASS*). In particular, the last measure allows us to draw key information about the presence of *inversion layers* at different heights from the ground, and therefore to investigate if some *hats* exist which prevent the pollutants to diffuse towards the higher layers of the atmosphere creating dangerous pollution phenomena.

The hypothesis is of course that the *meteorological predictions* are available, in such a way as to make simpler the network learning. Moreover the model has been taught to estimate the mean values of pollutant concentration. Such values are in fact considered by law to define air quality.

The input patterns for all the structures have been built in a similar way: in particular, as regards the *classifier*, its input consisted of the mean values for the meteorological measures as well as a certain number of delayed samples of the pollutant itself. Therefore the input pattern structure resembles a typical structure for a prediction model [5, 6], and indeed it does, but the prediction is performed only by associating the predicted value to two classes: *pre-alarm* condition or *normal* condition. Once predicted the situation, the same input pattern is processed by one out of the two different structures, able to predict the crisp value inside either a "pre-alarm" condition or a "normal" one. In particular, two different MLP structures perform such a task, depending on the class of membership for the particular pattern processed.

The whole predicting platform therefore reflects the scheme drawn in Figure15.1:

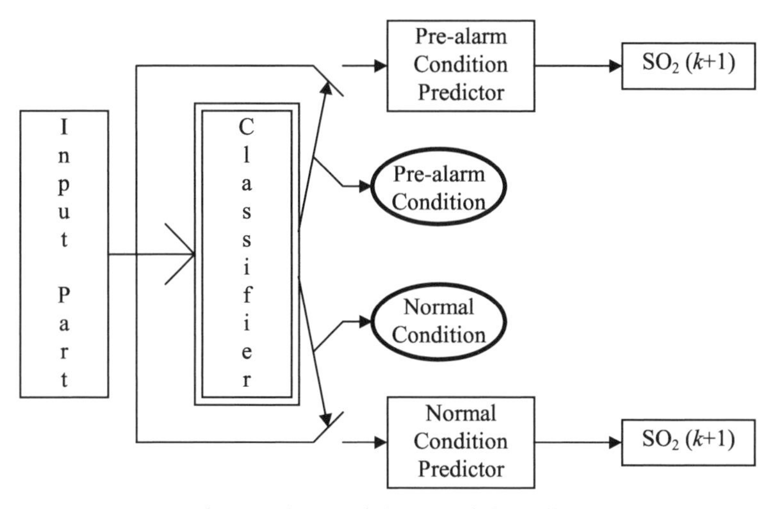

Figure 15.1. The neural network framework for pollutant estimation

15.5 The Model Structure Determination

The aim is to perform the pollutant prediction, and therefore to build-up a model of the phenomenon starting directly from data: the situation at hand obliges to perform a *black-box identification strategy* [3], *i.e.*, a methodology in which nothing is a-priori known about the model order, the number of parameter required for the modeling task, and so on.

In order to draw some information about the model structure, the auto-correlation function for the SO_2 pollutant time series has been considered is reported Figure15.2. The delayed samples mostly correlated with the current one should be considered to derive the model order.

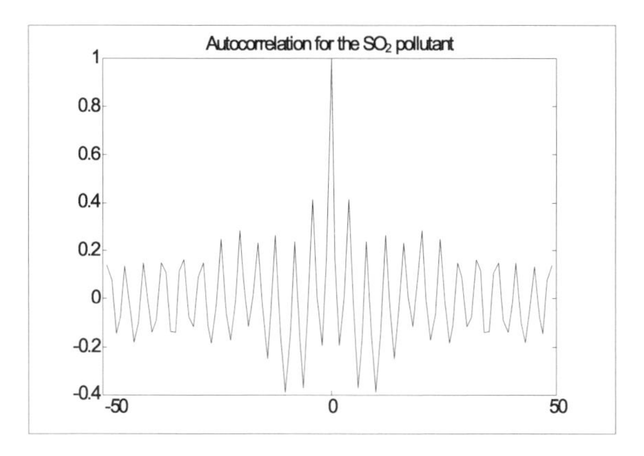

Figure 15.2. Auto-correlation function for the SO_2 time series considered

From the analysis of the *auto-correlation function* and after a few trials the model order has been fixed to 4

Following the same strategy, the *cross-correlation functions* between the SO_2 and the meteorological quantities time series have been considered in order to establish the delayed samples for each input signal needed for the model building-up.

The input pattern structure has therefore been fixed to:

$$[T(k+1)\ \ RH(k+1)\ \ WS(k)\ \ WS(k+1)\ \ WD(k+1)\ \ RASS(k-1)\ \ \mathrm{RASS}(k)\ \ RASS(k+1)\ RAD(k+1)\ P(k+1)\ R(k+1)\ SO_2(k-3)\ SO_2(k-2)\ SO_2(k-1)\ SO_2(k)]$$

The target is represented by: $[SO_2(k+1)]$.

The quantities present in the pattern refer, as previously mentioned, to the mean values of the measures taken in a 6-hours time interval, referring to data recorded at 30 minute time steps.

As regards the SO_2 samples for the input patterns of the classifier, they assumed the *low value* (0.2), *i.e.*, a *normal condition*, or the *high value* (0.8), *i.e.*, a *pre-alarm* condition, according to their class of membership. The target pattern consisted, for the classifier, in 0.2 or 0.8 according to the class of membership of $SO_2(k+1)$. As regards the predictor networks, the SO_2 samples assumed their real values in $\mu g/m^3$ of the level of the SO_2 sample to be predicted, while it assumed its real value in $\mu g/m^3$ for each of the predictors.

The data used to perform the whole work were drawn from a period of the year going from April to August 1992: such a choice because in this period the major pollution problems take place.

During the processing of such data several problems have been encountered, mainly due to the lack of measures. Some measures were absent at regular time periods, for the auto-calibration of the stations, but lots of other *practical* problems led to the lack of a considerable quantity of data. Since the mean SO_2 concentration was to be predicted, it was decided to perform 6-hours mean value prediction: due to the lack of data, the information was assumed to be valid only if the 6-hours window would contain at least 4 valid data. Otherwise the whole pattern was discarded. Such a *data filtering* heavily reduced the number of patterns available for the network learning and testing phase.

15.6 The "Classifier" Network

For the classifier learning, a threshold, fixed to about $65\mu g/m^3$ has been selected to separate the *pre-alarm* condition from the *normal* one. The network architecture has been fixed to the following structure:

15 input units; 10 hidden units; 1 output unit.

The number of hidden units has been derived with a *pruning strategy* implemented by using a graphic approach reported in [8]. Moreover the learning rate and the momentum term were tuned according to a procedure reported in [9]. A basic number of 150 patterns has been selected for the network training. Once reached the convergence, the following strategy is performed:

1. classify the following sample;
2. once measured its real value build the corresponding pattern and insert it in the learning set. Start a new learning phase and stop until the global error reaches the level of the previous learning phase, or until a fixed number of learning cycles is performed;
3. repeat Steps 1 and 2 for any sample.

The results of the classification step on a window not including the basic learning phase are reported in Figures 15.3, and 15.4.

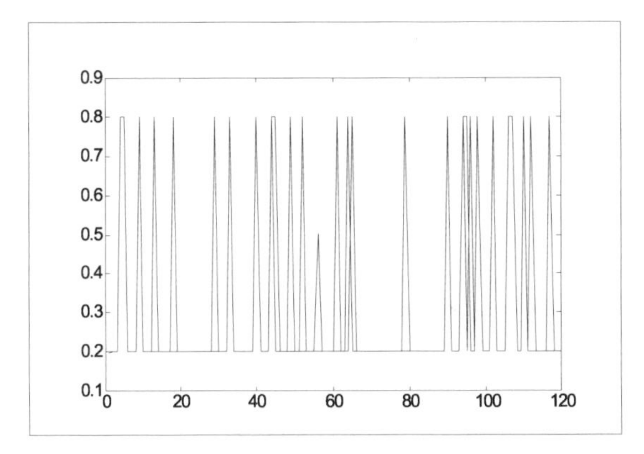

Figure 15.3. Normalized output for the "classifying" network over the test patterns

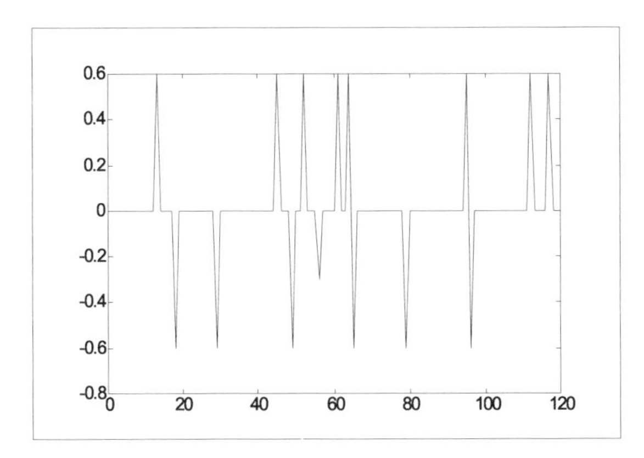

Figure 15.4. Error for the "classifying" network

The analysis of the figure reveals an error of about the 12%; but about the 6% reflects patterns which belong to an "indecision strip" across the threshold value of 65μg/m^3. To cope with such a problem the predictors were trained considering ranges of values of concentration partially overlapped with respect to the threshold: from 0 to 90μg/m^3 for the "normal condition", and from 22.5 to 270μg/m^3 for the "pre-alarm condition". In such a way the errors in the classifying due to uncertain patterns can be easily overcome, thus reducing the error to about 6%.

The response of the classifier simply acts as a switch that enables one out of the two predictors able to carry-on the estimation of the crisp value of the 6-hours ahead SO_2 mean value in each of the situations depicted by the classifier.

The patterns whose output is placed in the strip between 0.4 and 0.6 are considered as *not classified*. Such patterns reflect an *indecision* of the classifier, and their membership class remains unknown. They are processed by a different MLP network trained over a window containing 100 preceding patterns without no distinction between *pre-alarm* or *normal* condition. Such a network has to estimate the crisp value of the pollutant and therefore it represents the prediction step for the patterns that are *not classified*.

15.7 The "Prediction" Step

The fact that the number of patterns corresponding to the *pre-alarm* condition was smaller with respect to those belonging to the other class was derived from the patterns built: therefore the basic pattern number adopted for the predictor of the *pre-alarm* SO_2 mean value was 50, while the other one was of 100.

The prediction network learning and testing phase has been performed following the guidelines already depicted for the classification step. Moreover the following architecture for the two predictors has been used:

15 input units; 5 hidden units; 1 output unit.

The whole stage of classification-prediction is represented in Figures.15.5 and Figure 15.6.

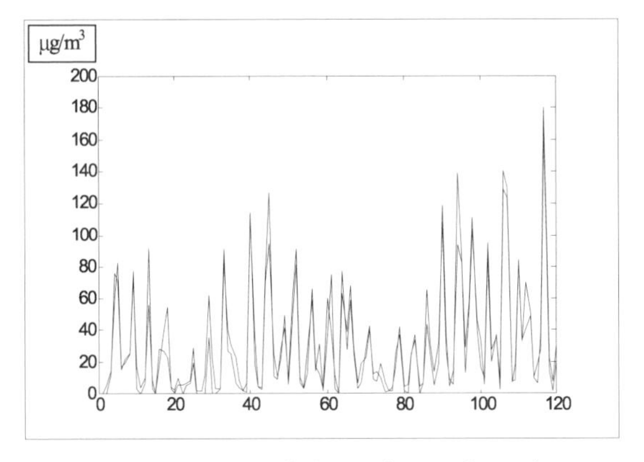

Figure 15.5. Classification-prediction task over the testing patterns

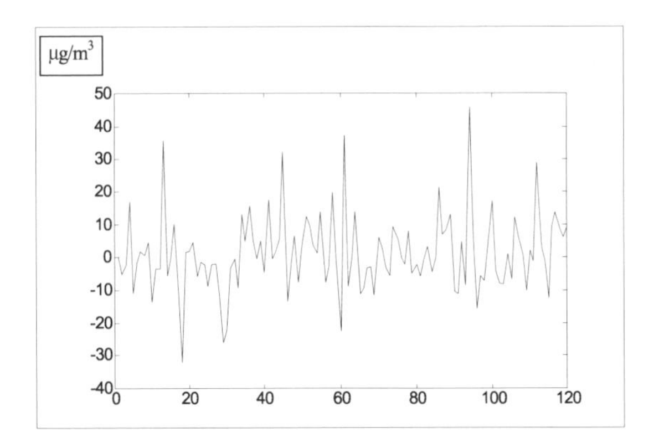

Figure 15.6. Error in the testing phase for the classification-prediction task

As it can be derived from the analysis of the figures, the estimation of the SO_2 mean value concentration is fairly good, notwithstanding the lack of a considerable quantity of measures. It is to be outlined that the neural architectures employed in both phases (classification and prediction) are small, and therefore they are suitable to be trained in *real-time*, allowing in a few minutes the SO_2 mean value estimation 6 hours ahead.

However, from the comparison between Figure 15.4 and Figure 15.6 it can be said that the higher errors values in the prediction step correspond to errors occurred in the *classification* step. The *classifier* remains therefore the most critical part in the whole prediction platform. Currently a considerable effort is being performed to find other strategies of learning able to provide better performance.

15.8 Conclusions

A novel approach, based on a neural platform, in order to perform the short-term (6-hours ahead) estimation of the mean value concentration of the SO_2 pollutant is reported. In particular, a *classification* step has been set-up to know in real-time if the current and estimated meteorological pollution conditions will lead to a *pre-alarm* situation or otherwise. Such a step represents a *rough prediction*, but it is very useful for several tasks, for example to make easier to work for the predictors (which have to predict mean value concentration inside limited ranges), to provide precious information to the petrochemical plants in order to plan the production policy with considerable economical advantages.

The results obtained can be considered satisfying, taking into account the enormous lack of measures, and above all the fact that such phenomena are highly complex and a systematic strategy for pollution forecasting does not exist. Moreover, the use of soft computing strategies to predict the concentration of air pollutants in both urban and industrial areas, not only for SO_2 but also for other common pollutants such as O_3, NO_2, NO_x, NMHC, has been proposed recently [10].

15.9 References

1. Sarokin, Schulkin. The Role of Pollution in Large-scale Population Disturbances. Part 2: Terrestrial Population Environmental Science. 1992; 26: 9
2. Alterio, Pignato, Zerbo. Multivariate Stochastic Model for Forecasting Hourly Concentrations of Air Pollutants in Melilli Area. Proc. IUAPPA Congress. 1993
3. Lyung, Soderstrom. Theory and Practice of Recursive Identification, Prentice Hall 1983
4. Arena P, Fortuna L, Gallo A, Nunnari G, Xibilia MG. Air Pollution Estimation via Neural Networks. Proc. 7th IMACS-IFAC Symp. on Large Scale Systems Theory and Applications (LSS'95), London UK, 1995
5. Cibenko G. Approximation by Superposition of a Ssigmoidal Function. Mathematics for Control, Signal and Systems. 1989; 2: 303-314
6. Chen, Billings, Grant. Non-linear System Identification Using Neural Networks. Int. J. Control, 1990; 51: (6): 1191-1214
7. Narendra, Parthasarathy. Identification and Control of Dynamical Systems Using Neural Networks. IEEE Trans. on Neural Networks. 1990; 1: 4-26
8. Arena P, Fortuna L, Graziani G, Nunnari G. A monitoring Approach for the design of Multilayer Neural Networks. Proc. COMADEM, Southampton, UK, 1991
9. Arena P, Fortuna L, Graziani S, Muscato G. A Real Time Implementation of a Multi-layer Perceptron with Automatic Tuning of Learning Parameters. IFAC Workshop on Algorithms and Architectures for Real Time Control, Bangor, North-Wales, Sept. 1991
10. Nunnari G, Nucifora A, Randieri C, (1998) The Application of Neural Techniques to the Modelling of Time Series of Atmospheric Pollution Data. Ecological Modelling, Vol. 111: 187-205.

16. Conclusions

16.1 Final Comments

In 1992, Lotfi Zadeh, the father of fuzzy logic, formalized in a few lines the objectives that soft computing techniques should pursue. His intuition that integration between neural networks, computational systems based on fuzzy logic, and optimization strategies inspired by evolutionist criteria should form the basis for rapid problem solving, which would have required considerable computational and algorithmic efforts using traditional-type methods, was fully confirmed.

This statement is justified by the considerable amount of applications based on soft computing techniques that have been developed over the last five years. Moreover, from 1992 to today, intense efforts of a methodological kind have allowed us, from the formal standpoint, to characterize those aspects of the basic techniques of soft computing which were hitherto underestimated.

We refer to the universal approximation theorems which have been proved both for neural networks and for fuzzy systems, and to genetic algorithm convergence analyses. Meanwhile, soft computing techniques today constitute a formally well-proven computational strategy.

The role played by soft computing has been fundamental in stimulating researchers who work in the information sector to approach other disciplines and to master those basic aspects from which non-conventional algorithms and innovative methodologies for solving complex problems in uncertain environments might be obtained.

Therefore, almost inadvertently, the birth of soft computing has coincided with an opening up towards activities in research fields apparently different from those initially pointed to by Zadeh.

About six years after Zadeh's definition, we find further tools that can be integrated with the traditional soft computing techniques, *i.e.*, chaotic dynamics, cellular neural networks (CNN), and the use of chaotic dynamic systems as associative memories.

Many of the most interesting methodological results have been formalized by reference to dynamic systems theory. The detailed study of complex dynamics has, moreover, led us to consider the possibility of using chaos as an active element in processing information, in such way that non-linear dynamics are completely integrated in soft computing techniques.

Furthermore, the evolution of soft computing strategies has kept pace with that of CNN [1], both from the architectural standpoint and from the applicational and circuital ones.

With regard particular to the last-mentioned aspect, the introduction of the CNN universal machine has been fundamental in the writing of this book. In Table 16.1, which will then be discussed below, the most salient interactions in the field of soft computing of the various tools essentially making it up, are schematically indicated with a view to synthesizing the concepts outlined in the text.

Let us take, for example, the interaction between fuzzy systems and cellular neural networks. The cellular neural network paradigm has been reproposed considering as a cellular element an elementary fuzzy system that allows us to model the equivalent circuital model of the cell in a traditional-type cellular neural network. And thus the concept of fuzzy CNN is born. This structure moreover allows us to model chaotic dynamics, which represents a further link between fuzzy CNN and chaotic dynamics. Furthermore, the definition of fuzzy CNN membership functions needs an optimization procedure that has to be solved by a global optimization algorithm; therefore, there evidently exists a degree of interaction between fuzzy CNN and genetic algorithms.

Table 16.1. Interactions among soft computing methodologies

	NEURAL NETWORKS (NN)	FUZZY SYSTEMS (FS)	GLOBAL OPTIMIZATION ALGORITHMS (GO)	CHAOTIC DYNAMICS (CD)	CELLULAR NEURAL NETWORKS (CNN)
NEURAL NETWORKS	Auxiliary neural networks for training	Neuro-fuzzy networks		NN needed for CD generation	Learning algorithms for CNN
FUZZY SYSTEMS	FS for NN auto-tuning		FS for choosing GA parameters	Generating CD with FS	FS definition of templates
GLOBAL OPTIMIZATION ALGORITHMS	GO for learning NN	GO for determining FS			Determining templates by GO
CHAOTIC DYNAMICS			CD as operators in GO		CNN with CD
CELLULAR NEURAL NETWORKS		Fuzzy CNN		Generating CD by CNN	

Bearing in mind that chaotic dynamics are fundamental for generating genetic algorithm mutations, a further link is evident between fuzzy CNN and global optimization algorithms. Very many other links can be discovered from a perusal of the above table.

From an architectural standpoint, many of the soft computing techniques in problem solving have a "distributed" type of calculation procedure rather than a "concentrated" one. The considerable success in modeling and controlling biological systems, made possible by the application of soft computing techniques in that field, deserves significant consideration.

Together with this, what can be noted in the field of mobile robotics is the ever more frequent design of intelligent robots, drawing inspiration in their mechanical structure from the world of animals (hexapods, molluscs, and fish).

It is thus natural to think of using soft computing techniques for the movements of such types of systems and for interpreting the scenario in which they operate.

Moreover, in the thematic field of artificial life and in emulating co-operative agents inspired by biological models, soft computing techniques allow us to generate efficient adaptive-type control strategies.

As is well known, for some decades now, the study being devoted to "complex" systems made up of a multiplicity of interconnected "simple" non-linear systems has been intense. The analytical kind of difficulties in handling them impose the use of approximation techniques, which are often insufficient for thoroughly characterizing their behavior; in addition, the chance of modeling them exclusively by means of experimental data would require considerable computational effort. Thus, the new idea consists in integrating analytical, heuristic, and experimental kinds of knowledge in order to avoid the preexistent limitations. Could soft computing be the suitable tool for analyzing and controlling such systems?

However, at the same time we should not overestimate the potential of these strategies that are used only when the traditional kind of techniques cannot be applied.

16.2 References

1. Arena P, Fortuna L, Manganaro G. Cellular Neural Networks: Chaos, Complexity and Vlsi Processing. Springer Verlag, 1999.
2. Arena P, Fortuna L, Branciforte M, Reaction-Diffusion CNN Algorithms to Generate and Control Artificial Locomotion. IEEE Transaction On Circuits and Systems, Vol. 1999; 46: (2): 253-260.

Index